T0205744

Universitext

Springer
New York
Berlin
Heidelberg
Barcelona
Budapest
Hong Kong
London
Milan
Paris
Santa Clara
Singapore
Tokyo

Morton L. Curtis

Matrix Groups

Second Edition

Morton L. Curtis

Rice University
Weiss School of Natural Sciences
Department of Mathematics
P.O. Box 1892
Houston, Texas 77001
USA

MSC 1991 20-01 20G99 22E20

Library of Congress Cataloging in Publication Data
Curtis, Morton Landers
Matrix groups.
(Universitext)
Bibliography: p.
Includes index.
1. Matrix groups. I. Title.
QA171.C87 1984 512'.2 84-14145

With 10 Illustrations

Printed and bound by Edwards Brothers, Inc., Ann Arbor, MI.
Printed in the United States of America.

9 8 7 6 5 4 3

ISBN 0-387-96074-0 Springer-Verlag New York Berlin Heidelberg
ISBN 3-540-96074-0 Springer-Verlag Berlin Heidelberg New York SPIN 10677223

To my teacher and friend
Raymond Louis Wilder
this book is affectionately dedicated.

Introduction

These notes were developed from a course taught at Rice University in the spring of 1976 and again at the University of Hawaii in the spring of 1977. It is assumed that the students know some linear algebra and a little about differentiation of vector-valued functions. The idea is to introduce students to some of the concepts of Lie group theory-- all done at the concrete level of matrix groups. As much as we could, we motivated developments as a means of deciding when two matrix groups (with different definitions) are isomorphic.

In Chapter I "group" is defined and examples are given; homomorphism and isomorphism are defined. For a field k , $M_n(k)$ denotes the algebra of $n \times n$ matrices over k . We recall that $A \in M_n(k)$ has an inverse if and only if $\det A \neq 0$, and define the general linear group $GL(n,k)$. We construct the skew-field $\mathbb{H}$ of quaternions and note that for $A \in M_n(\mathbb{H})$ to operate linearly on $\mathbb{H}^n$ we must operate on the right (since we multiply a vector by a scalar on the left). So we use row vectors for $\mathbb{R}^n$, $\mathbb{C}^n$, $\mathbb{H}^n$ and write xA for the row vector obtained by matrix multiplication. We get a <u>complex-valued</u> determinant function on $M_n(\mathbb{H})$ such that $\det A \neq 0$ guarantees that A has an inverse.

Chapter II introduces conjugation on $\mathbb{R} \subset \mathbb{C} \subset \mathbb{H}$ and then an inner product $\langle\ ,\ \rangle$. Basic properties of $\langle\ ,\ \rangle$ are given and then for $k \in \{\mathbb{R},\mathbb{C},\mathbb{H}\}$ we define the orthogonal group

$$\mathbb{O}(n,k) = \{A \in M_n(k) \mid \langle xA,yA\rangle = \langle x,y\rangle \text{ for all } x,y \in k^n\} .$$

$\mathfrak{O}(n,\mathbb{R})$ is written $\mathfrak{O}(n)$ and called the orthogonal group. $\mathfrak{O}(n,\mathbb{C})$ is written $U(n)$ and called the unitary group. $\mathfrak{O}(n,\mathbb{H})$ is written $Sp(n)$ and called the symplectic group. If $A \in \mathfrak{O}(n)$ then $\det A \in \{1,-1\}$ and the subgroup with $\det = 1$ is denoted by $SO(n)$ and called the special orthogonal group. If $A \in U(n)$ then $\det A$ is a complex number of unit length. The subgroup with $\det = 1$ is denoted by $SU(n)$ and called the special unitary group. As a first example of a matrix group isomorphism we show that $Sp(1) \cong SU(2)$.

In Chapter III we define the first invariant (i.e., something unchanged by an isomorphism) of a matrix group; namely, its dimension. A tangent vector to a matrix group G is $\gamma'(0)$ for some differentiable curve γ in G with $\gamma(0) = I$. The set T_G of all tangent vectors is shown to be a vector space, a real subspace of $M_n(k)$ ($k \in \{\mathbb{R},\mathbb{C},\mathbb{H}\}$). The dimension of T_G (as a real vector space) is the dimension of G . Smooth homomorphisms are defined and shown to induce linear maps of tangent spaces. Then dimension is seen to be an invariant.

In order, in Chapter IV, to calculate the dimensions of our matrix groups we develop the exponential map $\exp: M_n(k) \to GL(n,k)$ and the logarithm, $\log: U \to M_n(k)$ where U in some neighborhood of I in $GL(n,k)$. We have that exp and log are inverses, $\exp: V \to U$ and $\log: U \to V$ where V is a neighborhood of 0 on $M_n(k)$ and U is a neighborhood of I in $M_n(k)$ (actually $U \subset GL(n,k)$). One-parameter subgroups are defined and proved to be determined by their derivatives at 0 . It follows that T_G can be taken to be all derivatives of one-parameter subgroups. Lie algebras are defined and we see that each T_G is a Lie algebra. Finally, we then calculate the dimensions of $SO(n)$, $U(n)$, $SU(n)$ and $Sp(n)$.

In Chapter V we consider the very specific question of whether $Sp(1)$ and $SO(3)$ are isomorphic. We get a surjective homomorphism $\rho: Sp(1) \to SO(3)$ with kernel $= \{1,-1\}$. Then we define the center of a group, show it is an invariant and then calculate Center $Sp(1)$ $= \{1,-1\}$ and Center $SO(3) = \{I\}$, proving that $Sp(1) \not\cong SO(3)$. We define quotient groups and then note that we get new groups $\frac{G}{\text{center}}$ whenever a matrix group has nontrivial center.

In Chapter VI we do some topology which is needed in other parts of the text. All of our matrix groups are in some euclidean space and we just do topology of subsets of euclidean spaces. We give some basic results about continuity of functions, connected sets and compact sets. The proof that continuous functions preserve compactness relegated to an appendix. We consider countable bases for open sets since this is needed later in our study of maximal tori in matrix groups. Finally, there is a short section on manifolds.

Chapters VII, VIII, and IX are devoted to studying maximal tori in our matrix groups. We describe certain specific maximal tori. We prove that any two maximal tori are conjugate and that, if G is connected, then these conjugates cover G . At this stage we then know the dimension, center and rank of all of our matrix groups and these suffice to settle our original question as to which of these groups are isomorphic. At the end of Chapter IX we discuss simple groups and covering groups. The only new groups which arise are the double covers of $SO(n)$ $(n = 3,4,\ldots)$. This leads to the question:

Is the double cover of $SO(2n+1)$ isomorphic with $Sp(n)$?

In Chapter X we construct the double cover $Spin(n)$ of $SO(n)$ using Clifford algebras. We show that $Spin(1) \cong S^0$, $Spin(2) \cong S^1$, $Spin(3) \cong Sp(1)$ $(= S^3)$ and $Spin(4) \cong Sp(1) \times Sp(1)$. Finally we show that

$$Spin(5) \cong Sp(2) \quad \text{and} \quad Spin(6) \cong SU(4) .$$

In Chapter XI we finish our job by showing that $Sp(n) \not\cong Spin(2n+1)$ for any $n > 2$. This is done by looking at normalizers of maximal tori and resulting Weyl groups. If the normalizer is the semidirect product of the torus and the Weyl group, we say that the normalizer <u>splits</u>. If $Spin(2n+1)$ and $Sp(n)$ were isomorphic we would have

$$\frac{Spin(2n+1)}{center} \cong \frac{Sp(n)}{center} .$$

So our result is a consequence of the following three results.

(*) The normalizer in $Sp(n)$ does not split for any n .

(†) The normalizer in $\frac{Sp(n)}{center}$ splits $\Leftrightarrow n \in \{1,2\}$.

(Υ) The normalizer in $\frac{Spin(2n+1)}{center}$ $(= SO(2n+1))$ splits for $n = 1,2,3,\ldots$.

Finally, in Chapter XII we give a brief introduction to abstract Lie groups.

Introduction to the Second Edition

I have taken this occasion to add a short chapter introducing the concept of the roots of a compact connected Lie group. But mostly it has been an opportunity to correct a multitude of typos and some mathematical errors. In this regard I have been greatly helped by Bruce Sagan (University College of Wales) and by George Seligman (Yale). Both have sent me numerous helpful comments.

Morton L. Curtis
Houston, Texas
June 1984

Contents

Chapter 1 General Linear Groups 1

A. Groups 1
B. Fields, Quaternions 7
C. Vectors and Matrices 12
D. General Linear Groups 15
E. Exercises 20

Chapter 2 Orthogonal Groups 23

A. Inner Products 23
B. Orthogonal Groups 25
C. The Isomorphism Question 29
D. Reflections in $\mathbb{R}^n$ 31
E. Exercises 33

Chapter 3 Homomorphisms 35

A. Curves in a Vector Space 35
B. Smooth Homomorphisms 41
C. Exercises 43

Chapter 4 Exponential and Logarithm 45

A. Exponential of a Matrix 45
B. Logarithm 49
C. One-parameter Subgroups 51
D. Lie Algebras 56
E. Exercises 59

Chapter 5 SO(3) and Sp(1) 60

A. The Homomorphism $\rho : S^3 \to SO(3)$ 60
B. Centers 64
C. Quotient Groups 67
D. Exercises 71

Chapter 6 Topology 73

A. Introduction 73
B. Continuity of Functions, Open Sets, Closed Sets 74
C. Connected Sets, Compact Sets 79
D. Subspace Topology, Countable Bases 82
E. Manifolds 86
F. Exercises 89

Chapter 7 Maximal Tori 92

A. Cartesian Products of Groups 92
B. Maximal Tori in Groups 95
C. Centers Again 100
D. Exercises 104

Chapter 8 Covering by Maximal Tori 106

A. General Remarks 106
B. (+) for U(n) and SU(n) 108
C. (+) for SO(n) 111
D. (+) for Sp(n) 116
E. Reflections in R^n (again) 117
F. Exercises 120

Chapter 9 Conjugacy of Maximal Tori 122

A. Monogenic Groups 122
B. Conjugacy of Maximal Tori 124
C. The Isomorphism Question Again 125
D. Simple Groups, Simply-Connected Groups 127
E. Exercises 130

Chapter 10 Spin(k) 131

A. Clifford Algebras 131
B. Pin(k) and Spin(k) 135
C. The Isomorphisms 140
D. Exercises 142

Chapter 11 Normalizers, Weyl Groups 143

A. Normalizers 143
B. Weyl Groups 147
C. Spin(2n+1) and Sp(n) 149
D. SO(n) Splits 154
E. Exercises 159

Chapter 12 Lie Groups 161

A. Differentiable Manifolds 161
B. Tangent Vectors, Vector Fields 162
C. Lie Groups 170
D. Connected Groups 175
E. Abelian Groups 180

Chapter 13 182

A. Maximal Tori 182
B. The Anatomy of a Reflection 187
C. The Adjoint Representation 190
D. Sample Computation of Roots 196

Appendix 1 201
Appendix 2 203
References 204
Index 205
Supplementary Index (for Chapter 13) 210

Chapter 1
General Linear Groups

A. Groups

Before we can discuss matrix groups we need to talk a little about groups in general. If X and Y are sets, their Cartesian product $X \times Y$ is defined to be the set of all ordered pairs (x,y) with $x \in X$ and $y \in Y$. A convenient notation for describing this set of all ordered pairs is

$$X \times Y = \{(x,y) | x \in X \text{ and } y \in Y\} ,$$

the curly brackets being read as "the set of all" and the vertical bar as "such that."

By a binary operation ρ on a set S we mean a function

$$\rho : S \times S \to S ;$$

i.e., for an ordered pair (s_1,s_2) of elements of S, ρ assigns another element of S which we write as $\rho(s_1,s_2)$. For example, the set $N = \{1,2,3,...\}$ of natural numbers has two well-known binary operations on it. Addition sends the ordered pair (a,b) of natural numbers to the natural number $a + b$. Multiplication sends the ordered pair (a,b) to ab.

<u>Definition</u>: A <u>group</u> G is a set G along with a binary operation

$$\phi : G \times G \to G$$

satisfying certain properties. To state these properties it is convenient to adopt a simple notation--for $\phi(a,b)$ we just write ab .

Required properties of the operation:

(i) The operation is <u>associative</u>. This means that for any $a,b,c \in G$ we have

$$(ab)c = a(bc) \ .$$

(If we had maintained the $\phi(a,b)$ notation this would read $\phi(\phi(a,b),c) = \phi(a,\phi(b,c))$.

(ii) There exists an <u>identity</u> element e of G . This means that for any $a \in G$ we have $ea = ae = a$.

(iii) <u>Inverses</u> exist. This means that for any $a \in G$ there is an element $a^{-1} \in G$ such that $aa^{-1} = a^{-1}a = e$.

Note that properties (ii) and (iii) leave open the possibilities that there may be more than one identity element and that an element may have more than one inverse. But neither of these can happen.

<u>Proposition</u> 1: <u>A group G has exactly one identity element and each $a \in G$ has exactly one inverse.</u>

<u>Proof</u>: Suppose e and f are identity elements of G . Then

$$fe = e \quad \text{since} \quad f \quad \text{is an identity element, and}$$

$$fe = f \quad \text{since} \quad e \quad \text{is an identity element .}$$

Suppose both b and c are inverses of a . Then

$$b = eb = (ca)b = c(ab) = ce = c .$$

Examples

(1) The set $\mathbb{Z} = \{\ldots,-2,-1,0,1,2,\ldots\}$ of integers is a group under addition. 0 is the identity and the inverse of a is $-a$.

(2) $\mathbb{Z}$ is not a group under multiplication. The operation is associative and 1 is the identity. But, for example, there is no inverse for 2 .

(3) The set $\mathbb{Q}$ of rational numbers is a group under addition.

(4) The set $\mathbb{Q} - (0)$ (i.e., all nonzero rationals) is a group under multiplication.

(5) $\mathbb{R}^+ = \{x \in \mathbb{R} | x > 0\}$ is the set of all positive real numbers. It forms a group under multiplication.

(6) $\mathbb{R}^n$ = the set of all ordered n-tuples of real numbers is a group under the following operation: if

$$x = (x_1,x_2,\ldots,x_n) \quad \text{and}$$

$$y = (y_1,y_2,\ldots,y_n), \quad \text{then}$$

$$x + y = (x_1 + y_1,x_2 + y_2,\ldots,x_n + y_n) \quad .$$

The identity is

$$\Theta = (0,0,\ldots,0)$$

and the inverse of x is $(-x_1,-x_2,\ldots,-x_n)$.

(7) Let $S = \{a,b,c\}$; i.e., S is a set with three elements which we denote by a,b,c . Let G be the set of all one-to-one maps (functions) of S onto S . For example $f : S \to S$ given by $f(a) = b$, $f(b) = c$, $f(c) = a$ is one element of G . We define an <u>operation</u> on G as follows: if $f,g \in G$ we let

$$f \circ g : S \to S$$

be defined by $(f\circ g)(a) = f(g(a))$, $(f\circ g(b) = f(g(b))$, $(f\circ g)(c) = f(g(c))$, i.e., $f\circ g$ means first apply g to S and then apply f. Let $i : S \to S$ be the identity element $(i(a) = a,\ i(b) = b,\ i(c) = c)$. Then this is the identity element for G for this operation. Then the usual inverse of $f \in G$ is the inverse for f relative to this operation. Thus G is a group. It is called the <u>symmetric group</u> on $\{a,b,c\}$ (or just the symmetric group on three elements).

<u>Definition</u>: A group G is <u>abelian</u> if for every $a,b \in G$ we have $ab = ba$.

In the examples above, (1), (3), (4), (5), and (6) are abelian groups, but the symmetric group on three elements is not abelian. (Exercise.)

The kind of functions (mapping one group to another) of interest to us are those which "preserve" the operations--these are called <u>homomorphisms</u>.

<u>Definition</u>: Let G and H be groups. A function $\sigma : G \to H$ is

a homomorphism if for every a,b in G we have

$$\sigma(ab) = \sigma(a)\sigma(b) .$$

What this means is that we can first multiply a and b (using the operation in G) and then map the result by σ, or we can map a and b into H by σ and multiply there--with the same result.

Proposition 2: *A homomorphism* $\sigma : G \to H$ *sends identity to identity and inverses to inverses.*

Proof: Let e,e' be the identities in G,H. We have $\sigma(e) = \sigma(ee) = \sigma(e)\sigma(e)$ and $\sigma(e)$ has an inverse, call it h, in H. So

$$e' = h\sigma(e) = h\sigma(e)\sigma(e) = \sigma(e) .$$

For $a \in G$ we have

$$\sigma(a)\sigma(a^{-1}) = \sigma(aa^{-1}) = \sigma(e) = e' ,$$

showing that $\sigma(a^{-1}) = (\sigma(a))^{-1}$.

A homomorphism is surjective (or onto) if $\sigma(G) = H$. If we define $\sigma : \mathbb{R} \to \mathbb{R}^2$ ($\mathbb{R}$ = additive group of reals, $\mathbb{R}^2$ as in example (6)) by $\sigma(x) = (x,x)$, then σ is a homomorphism but is not surjective because $\sigma(\mathbb{R})$ is just the diagonal line in $\mathbb{R}^2$. But $\rho : \mathbb{R}^2 \to \mathbb{R}$ defined by $\rho(x,y) = x$ is a surjective homomorphism.

A homomorphism $\sigma : G \to H$ is injective if $\sigma(a) = \sigma(b)$ always implies $a = b$; i.e., no two elements go to the same place. Sometimes this is called one-to-one-into, but we won't do that. For example, the map $\sigma : \mathbb{R} \to \mathbb{R}^2$ ($\sigma(x) = (x,x)$) is injective, and the map

$\rho : R^2 \to R$ $(\rho(x,y) = x)$ is not injective.

A homomorphism which is both injective and surjective is called an <u>isomorphism</u>. From an abstract point of view, two groups which are isomorphic are "really" the same group--even if they were defined in strikingly different manners. There is a classic example of this.

Let R be the additive group of all real numbers and let R^+ (see Example 5) be the multiplicative group of all positive real numbers. Let a be any real number greater than 1. Define

$$\sigma : R \to R^+$$

by

$$\sigma(x) = a^x .$$

Then σ is a homomorphism

$$\sigma(x + y) = a^{x+y} = a^x a^y = \sigma(x)\sigma(y) .$$

Also, σ is injective. For, suppose $\sigma(x) = \sigma(y)$. This means $a^x = a^y$ and so $a^{-y}a^x = a^{-y}a^y = 1$ and $a^{x-y} = 1$ which implies $x - y = 0$ or $x = y$. Also, σ is surjective. For, if y is any positive real number $x = \log_a y$ has the property that $a^x = y$. Thus these two groups are isomorphic--not only that, but there are lots of isomorphisms.

We conclude this section with a simple, but important, remark. A priori it looks difficult to see if a homomorphism $\sigma : G \to H$ is injective. Do we really have to check all pairs a,b in G to see if $\sigma(a) = \sigma(b)$? Fortunately not.

$$\sigma \text{ is injective} \Leftrightarrow \sigma^{-1}(e') = e .$$

$$\sigma(a) = \sigma(b) \Leftrightarrow \sigma(a)\sigma(b)^{-1} = e' \Leftrightarrow \sigma(ab^{-1}) = e'$$

and

$$ab^{-1} = e \Leftrightarrow a = b .$$

B. Fields, Quaternions

Definition: A field k is a set that has operations of addition and multiplication satisfying certain requirements:

(i) multiplication distributes over addition;

$$a(b + c) = ab + ac ;$$

(ii) k is an abelian group, with identity written as 0, under addition.

(iii) $k - (0)$ is an abelian group under multiplication.

Examples. The rationals $\mathbb{Q}$ and the reals $\mathbb{R}$ are fields. We can make $\mathbb{R}^2$ into a field $\mathbb{C}$ (the complex numbers) as follows. If (x_1,x_2) and (y_1,y_2) are two ordered pairs of real numbers, we define $(x_1,x_2) + (y_1,y_2) = (x_1 + y_1,x_2 + y_2)$ and we have seen that this operation makes $\mathbb{R}^2$ into an abelian group. Suppose for multiplication we try

$$(x_1.x_2)(y_1.y_2) = (x_1y_1,x_2y_2)$$

(surely the most obvious thing). Then we would have

$$(1,0)(0,1) = (0,0) .$$

Now $(0,0)$ is the additive identity or "zero" and we would have two nonzero elements of R^2 with a zero product. The result could not be a field because:

<u>Proposition</u> 3: <u>In a field</u> k <u>if</u> $a \neq 0$ <u>and</u> $b \neq 0$, <u>then</u>

$$ab \neq 0 .$$

<u>Proof</u>: If $a \neq 0$ then $a \in k - (0)$ which by (iii) is required to be a group under multiplication. Thus there is an a^{-1} in $k - (0)$ such that $a^{-1}a = 1$ (the multiplicative identity). Thus if $ab = 0$ we have

$$a^{-1}(ab) = (a^{-1})(0) = 0$$

but $a^{-1}(ab) = (a^{-1}a)b = 1b = 0$ so $b = 0$.

The statement of Proposition 3 is equivalent to the statement that a field has "<u>no divisors of zero</u>."

So how do we make R^2 into a field? Our most naive attempt failed flat. Well, what turns out to work is

$$(a,b)(c,d) = (ac - bd, ad + bc) .$$

We must first verify that this distributes over addition.

$$(a,b)((c,d) + (e,f)) = (a,b)((c + e, d + f))$$

$$= (a(c + e) - b(d + f), a(d + f) + b(c + e) .$$

This should equal $(a,b)(c,d) + (a,b)(e,f)$. This latter equals $(ac - bd, ad + bf) + (ad - bf, af + be)$ and we easily check that these are equal. Next we need to see that if $(a,b) \neq (0,0)$ then it has a multiplicative inverse. Well, $(a,b) \neq (0,0) \Leftrightarrow a^2 + b^2 \neq 0$; in which case, we need to find a multiplicative inverse for (a,b) . The multiplicative identity clearly is $(1,0)$ and

$$(a,b)\left(\frac{a}{a^2+b^2}, \frac{-b}{a^2+b^2}\right) = (1,0) ,$$

as you can readily verify. Thus we have made $\mathbb{R}^2$ into a field which we denote by $\mathbb{C}$ and call the <u>complex numbers</u>.

You may know that there is a simple mnemonic device for remembering multiplication in $\mathbb{C}$. Write $(a,b) = a + ib$ or $a + bi$ and treat these as polynomials in i with the side condition that $i^2 = -1$. Thus

$$\begin{aligned}(a + ib)(c + id) &= ac + aid + ibc + ibid \\ &= ac + iad + ibc + i^2bd \\ &= (ac - bd) + i(ad + bc) .\end{aligned}$$

We can consider $\mathbb{R}$ to be a subfield of $\mathbb{C}$ (i.e., a subset which becomes a field using the operations in the larger set) by letting

$$x \in \mathbb{R} \text{ be } x + i0 .$$

Then if $x,y \in \mathbb{R}$ we have

$$x + y = x + i0 + y + i0 = (x + y) + i0$$

$$xy = (x + i0)(Y + i0) = (xy) + i0 .$$

So we have taken the field $\mathbb{R}$ as all $(x,0)$ in $\mathbb{R}^2$ and <u>extended</u>

the operations in $\mathbb{R}$ to $\mathbb{R}^2$ to get a field.

This strongly suggests that we try to extend the field on $\mathbb{R}^2$ to a field on $\mathbb{R}^3$. Now for the bad news.

Proposition 4: The operations on $\mathbb{C}$ cannot be extended to make $\mathbb{R}^3$ into a field.

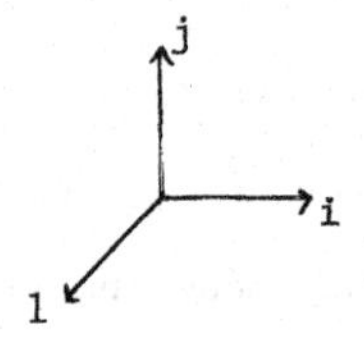

Proof: Take basis vectors $1, i, j$ so that any element of $\mathbb{R}^3$ can be written uniquely as $a + ib + jc$ with $a, b, c \in \mathbb{R}$. So if we are to have a multiplication extending that of $\mathbb{C}$ we must have $ij = a + ib + jc$ for some three real numbers a, b, c. But then $i(ij) = ia + i^2b + ijc$; so

$$-j = ia - b + ijc$$

$$-j = ia - b + (a + ib + jc)c$$

$$-j = (ac - b) + i(a + bc) + jc^2.$$

This implies $c^2 = -1$, contradicting $c \in \mathbb{R}$.

The main thrust of this proof is that if we insist that the product ij be in $\mathbb{R}^3$ we get into trouble. Maybe if we had one more dimension it would work. This is almost true; we can define a multiplcation on $\mathbb{R}^4$ which satisfies conditions (i) and (ii) for a field but (iii) must be replaced by (iii)′, $k - (0)$ is a group under multiplication--it is not an abelian group. We will just describe how this can be done. You may be interested in reading "Hamilton's discovery of the quaternions" by B. L. van der Waerden in the Mathematics Magazine (vol. 49, #5, (1976)). We take a basis $1, i, j, k$ for $\mathbb{R}^4$

and define

	1	i	j	k
1	1	i	j	k
i	i	-1	k	-j
j	j	-k	-1	i
k	k	j	-i	-1

Thus 1 acts as identity, $ij = k$, $ji = -k$, etc.

This tells us how to multiply quadruples of real numbers:

$$(a + ib + jc + kd)(x + iy + jz + kw) = (ax - by - cz - dw)$$

$$+ i(ay + bx + cw - dz) + j(az + cx + dy - bw)$$

$$+ k(aw + dx + bz - cy) .$$

$\mathbb{R}^4$ with this multiplication is called the quaternions. It is easy to verify that this does extend the multiplication in $\mathbb{C}$ by taking $c = 0 = d$ and $z = 0 = w$ in the formula above. The modified field axioms (i), (ii), (iii)′ are readily verified except for showing that every nonzero quaternion has an inverse. But if

$$q = a + ib + jc + bd$$

is not the zero $(0 + i0 + j0 + k0)$ then $a^2 + b^2 + c^2 + d^2 \neq 0$ and we set

$$q^{-1} = \frac{a - ib - jc - kd}{a^2+b^2+c^2+d^2} ,$$

and readily verify that $qq^{-1} = 1 = q^{-1}q$.

There are certain constructions we want to make for $\mathbb{R}$ and $\mathbb{C}$ and the quaternions (which we denote by $\mathbb{H}$), so we will write

$$k \in \{\mathbb{R},\mathbb{C},\mathbb{H}\} .$$

C. Vectors and Matrices

For $k \in \{\mathbb{R},\mathbb{C},\mathbb{H}\}$ let k^n be the set of all ordered n-tuples of elements of k. Define addition on k^n by

$$x = (x_1,\ldots,x_n) \qquad y = (y_1,\ldots,y_n)$$

$$x + y = (x_1 + y_1,\ldots,x_n + y_n) .$$

This makes k^n into an abelian group with identity $\mathbb{O} = (0,\ldots,0)$. For $c \in k$ we define

$$cx = (cx_1,\ldots,cx_n)$$

and this makes k^n into a vector space over k (for $k = \mathbb{H}$ we must relax the usual definition which insists that k be a field).

Definition: A map $k^n \xrightarrow{\phi} k^n$ is linear if it respects linear combinations; i.e., if $c,d \in k$ and $x,y \in k^n$ then

$$(*) \qquad \phi(cx + dy) = c\phi(x) + d\phi(y) .$$

In particular, $\phi(x + y) = \phi(x) + \phi(y)$, so that a linear map is a homomorphism of the additive group of k^n. Also

$$\phi(cx) = c\phi(x) ,$$

and these two conditions together are equivalent to (*).

Proposition 5: If $k^n \xrightarrow{\phi} k^n \xrightarrow{\psi} k^n$ are both linear, then

<u>so is</u> $\psi\circ\phi$.

<u>Proof</u>: $(\psi\circ\phi)(cx + dy) = \psi(c\phi(x) + d\phi(y))$

$$= c(\psi\circ\phi)(x) + d(\psi\circ\phi)(y) .$$

<u>Definition</u>: $M_n(k)$ is the set of all $n \times n$ matrices with elements from k .

If $M \in M_n(K)$, $M = (m_{ij})$ $(m_{ij} \in k)$, we can define a linear map $\phi(M)$ by

$$\phi(M)(x_1,\ldots,x_n) = (x_1,\ldots,x_n)(m_{ij})$$

where matrix multiplication is indicated on the right; i.e., we are multiplying a $1 \times n$ matrix with an $n \times n$ matrix to give a $1 \times n$ matrix. This is easily seen to be linear.

$$\phi(M)(cx + dy) = (cx + dy)\ (m_{ij})$$

$$= c(x_1,\ldots,x_n)(m_{ij}) + d(y_1,\ldots,y_n)(m_{ij}) .$$

We use row vectors instead of column vectors because <u>we no longer have a choice</u> when $k = \mathbb{H}$. We made $\mathbb{H}^n$ into a vector space by defining scalar multiplication <u>on the left</u>,

$$c(x_1,\ldots,x_n) = (cx_1,\ldots,cx_n)$$

and this is not the same as $(x_1c.\ldots.x_nc)$ in general. If we use column vectors and multiply by matrices on the left we <u>do not always get linear maps</u>. For $q,c,d \in \mathbb{H}$ and $x,y \in \mathbb{H}^n$ consider

$$\begin{pmatrix} q & & 0 \\ & \ddots & \\ 0 & & q \end{pmatrix} \begin{pmatrix} cx_1 + dy_1 \\ \vdots \\ cx_n + dy_n \end{pmatrix} = \begin{pmatrix} qcx_1 + qdy_1 \\ \vdots \\ qcx_n + qdy_n \end{pmatrix}$$

and we certainly can't expect this to equal

$$c\begin{pmatrix} qx_1 \\ \vdots \\ qx_n \end{pmatrix} + d\begin{pmatrix} qy_1 \\ \vdots \\ qy_n \end{pmatrix} .$$

(Take $n = 1$, $x = 1$, $y = 1$, $d = 0$, $c = i$ and $q = j$.)

Conversely, given a linear map $\phi : k^n \to k^n$, it is easy to find an $n \times n$ matrix M such that $\phi = \phi(M)$ (and it will clearly be unique). The first row of M is the n-tuple $\phi(1,0,\ldots,0)$, the second row of M is $\phi(0,1,0,\ldots,0)$, etc.

Note that if the matrix A gives the linear map ϕ and the matrix B gives the linear map ψ then AB gives $\psi \circ \phi$. A linear map ϕ is an <u>isomorphism</u> if it is injective and surjective (same definitions as for group homomorphisms). Then ϕ^{-1} is also a linear isomorphism and $\phi \circ \phi^{-1}$ = identity map = $\phi^{-1} \circ \phi$. For the corresponding matrices this means that $M(\phi^{-1})M(\phi) = I = M(\phi)M(\phi^{-1})$ so that $M(\phi^{-1})$ is a 2-sided inverse for $M(\phi)$. So <u>if</u> A^{-1} <u>is a left inverse for</u> A, <u>then it is also a right inverse for</u> A.

We make the set $M_n(k)$ into a vector space in a fairly obvious way:

(i) If $A = (a_{ij})$ and $B = (b_{ij})$, then

$$A + B = (a_{ij} + b_{ij}) ;$$

(ii) If $A = (a_{ij})$ and $c \in k$, then

$$cA = (ca_{ij}) .$$

This is really no different from the way we made k^n into a vector space, but we are now working with n^2-tuples. However, there is nothing to be gained by writing n^2 elements in a line instead of an $n \times n$ array.

But $M_n(k)$ is not just a vector space. It also has a multiplication which distributes over addition (on either side).

$$A(B + C) = AB + AC$$

$$(B + C)A = BA + CA .$$

Such a system is called an <u>algebra</u>. When we use the word algebra we will always mean one with a two-sided multiplicative identity. For $M_n(k)$,

$$I = \begin{pmatrix} 1_1 & & \bigcirc \\ & \ddots & \\ \bigcirc & & 1 \end{pmatrix}$$

is the multiplicative identity.

D. General Linear Groups

<u>Definition</u>: If $\mathcal{A}$ is an algebra, $x \in \mathcal{A}$ is a <u>unit</u> if there exists some $y \in \mathcal{A}$ such that $xy = 1 = yx$, i.e., if it has a multiplicative inverse.

<u>Proposition</u> 6: <u>If</u> $\mathcal{A}$ <u>is an algebra with an associative</u>

<u>multiplication</u> <u>and</u> $U \subset G$ <u>is the set of units in</u> G , <u>then</u> U <u>is a group under multiplication</u>.

<u>Proof</u>: The operation is associative, there is an identity element 1 and every element has an inverse.

<u>Definition</u>: The group of units in the algebra $M_n(\mathbb{R})$ is denoted by $GL(n,\mathbb{R})$, in $M_n(\mathbb{C})$ by $GL(n,\mathbb{C})$ and in $M_n(\mathbb{H})$ by $GL(n,\mathbb{H})$. These are the <u>general linear groups</u>.

Note that: $A \in M_n(k)$ is a unit $\Leftrightarrow$ A represents an isomorphism of k^n .

<u>Definition</u>: If G is a group and H is a subset of G , then H is a <u>subgroup</u> of G if the operation on G makes H into a group.

<u>Proposition</u> 7: H <u>is a subgroup of the group</u> G <u>if</u> ($H \subset G$ <u>and</u>)

(i) $x,y \in H \Rightarrow xy \in H$,

(ii) <u>id. el. is in</u> H ,

(iii) $x \in H \Rightarrow x^{-1} \in H$.

<u>Proof</u>: (Exercise.) The subject of this course is the study of subgroups of these general linear groups.

A 1×1 matrix over k is just an element of k and matrix multiplication of two is just multiplication in k . So we see that

$$GL(1,\mathbb{R}) = \mathbb{R} - (0)$$
$$GL(1,\mathbb{C}) = \mathbb{C} - (0)$$
$$GL(1,\mathbb{H}) = \mathbb{H} - (0)$$

because all nonzero elements are units. $GL(2,\mathbb{R})$ is the set of units in the vector space $M_2(\mathbb{R})$ of dimension 4 . So

$$GL(2,\mathbb{R}) = \{\begin{pmatrix} a & b \\ c & d \end{pmatrix} \mid \begin{matrix} a,b,c,d \in \mathbb{R} \\ ad - bc \neq 0 \end{matrix}\} ,$$

i.e., all points in 4-space not on the set where $ad = bc$.

For $\mathbb{R}$ and $\mathbb{C}$ we have determinants defined on $M_n(\mathbb{R})$ and $M_n(\mathbb{C})$ and from linear algebra we know that

$$GL(n,\mathbb{R}) = \{A \in M_n(\mathbb{R}) \mid \det A \neq 0\}$$

$$GL(n,\mathbb{C}) = \{A \in M_n(\mathbb{C}) \mid \det A \neq 0\} .$$

Suppose we define a "determinant" on $M_2(\mathbb{H})$ by

$$\det\begin{pmatrix} \alpha & \beta \\ \gamma & \delta \end{pmatrix} = \alpha\delta - \beta\gamma .$$

Then $\det\begin{pmatrix} i & j \\ i & j \end{pmatrix} = k - (-k) = 2k \neq 0$, but this matrix cannot be a unit or the corresponding linear map would be an isomorphism, whereas

$$(j,-j)\begin{pmatrix} i & j \\ i & j \end{pmatrix} = (0,0)$$

and the map is not injective. Similar definitions give similar problems, but we can define a <u>complex-valued</u> determinant with the desired property: namely, $A \in M_n(\mathbb{H})$ has an inverse if and only if this determinant is nonzero.

Proposition 8: *Let $\phi : G \to H$ be a homomorphism of groups. Then $\phi(G)$ is a subgroup of H.*

Proof: $\phi(id) = id$ so that $\phi(G)$ contains the identity element of H. If $x.y \in \phi(G)$ there exist $a,b \in G$ such that $\phi(a) = x$, $\phi(b) = y$. Then

$$xy = \phi(a)\phi(b) = \phi(ab) \in \phi(G) \ .$$

Finally, suppose $x \in \phi(G)$. Then $x = \phi(a)$ and so $x^{-1} = \phi(a^{-1})$ $\in \phi(G)$. So $\phi(G)$ is a subgroup of H .

If $\phi : G \to H$ is an injective homomorphism, then ϕ is an isomorphism of G onto the subgroup $\phi(G)$ of H , so we can then consider G as a subgroup of H . We are going to construct an injective homomorphism

$$\Psi : GL(n,\mathbb{H}) \to GL(2n,\mathbb{C}) \ ,$$

and then for $A \in GL(n,\mathbb{H})$ we will assign as the determinant of A the determinant of $\Psi(A)$.

We begin with

$$\psi : \mathbb{H} \to M_2(\mathbb{C})$$

defined by

$$\psi(x+iy+jz+kw) = \begin{pmatrix} x+iy & -z-iw \\ z-iw & x-iy \end{pmatrix} \ .$$

<u>Lemma</u> 9: (i) $\psi(\alpha+\beta) = \psi(\alpha) + \psi(\beta)$

(ii) $\psi(\alpha\beta) = \psi(\alpha)\psi(\beta)$

(iii) ψ <u>is injective</u> .

<u>Proof</u>: (i) is trivial and (ii) is a routine, but somewhat tedious, computation, and (iii) is trivial.

Next, for $A \in Mn(\mathbb{H})$ we set

$$\Psi(A) = (\psi(a_{ij}))$$

i.e. $\Psi(A)$ is the complex $2n \times 2n$ matrix whose 2×2 block in the ij position is $\psi(a_{ij})$.

Lemma 10: $\Psi(AB) = \Psi(A)\Psi(B)$.

Proof: Let $A = (\alpha_{uv})$, $B = (\beta_{uv})$. Then

$$(AB)_{ij} = \alpha_{i1}\beta_{1j}+\cdots+\alpha_{in}\beta_{nj} \text{ . By Lemma 9}$$

$$(\Psi(AB)_{ij}) = \psi(\alpha_{i1})\psi(\beta_{1j})+\cdots+\psi(\alpha_{in})\psi(\beta_{nj})$$

and this is just the ij entry in $\Psi(A)\Psi(B)$.

Proposition 9: If $A \in M_n(\mathbb{H})$, then $A \in GL(n,\mathbb{H}) \Leftrightarrow \det A \neq 0$.

Proof: $A \in GL(n,H)$ implies that there exists $A^{-1} \in GL(n,H)$ with $AA^{-1} = I = A^{-1}A$. Then $\psi(A)$ has $\psi(A^{-1})$ as its inverse (in $GL(2n,\mathbb{C})$) and thus $\det A = \det \psi((A)$ is nonzero. Now $GL(n,\mathbb{H})$ is the group of units in $M_n(\mathbb{H})$ and $\psi(GL(n,\mathbb{H})$ is the group of units in $\psi(M_n(\mathbb{H}))$. We need to show that $(\psi(A))^{-1}$ lies in $\psi(GL(n,\mathbb{H})$. This is a consequence of the following general result.

Proposition 10: Let $\mathcal{B}$ be a finite-dimensional associative algebra with 1. Let $\mathcal{A}$ be a subalgebra of $\mathcal{B}$ (with $1 \in \mathcal{A}$). If $U(\mathcal{A})$ and $U(\mathcal{B})$ denote the gropus of units, then

$$U(\mathcal{A}) = \mathcal{A} \cap U(\mathcal{B}).$$

Proof: The proof of this proposition is given in Appendix 2. It involves some concepts not yet introduced and would tend to disrupt what we are doing here.

E. Exercises

1. Let $\phi : G \to H$ be a homomorphism of groups. The <u>kernel</u> of ϕ is defined to be

$$\ker \phi = \{x \in G \mid \phi(x) = \text{identity of } H\} \quad .$$

Show that $\ker \phi$ is a subgroup of G .

2. A subgroup W of a group G is <u>normal</u> if for each $x \in G$ we have

$$xWx^{-1} = W \ .$$

Show that $\ker \phi$ (Exercise 1) is a normal subgroup of G .

3. The <u>center</u> C of a group G is defined by

$$C = \{y \in G \mid xy = yx \quad \text{for all} \quad x \in G\} \quad .$$

Show that C is a normal subgroup of G .

4. Let S be a nonempty subset in a group G . Define the <u>centralizer</u> $C(S)$ of S by

$$C(S) = \{x \in G \mid xs = sx \quad \text{for all} \quad s \in S\} \quad .$$

Show that $C(S)$ is a subgroup of G .

5. Let S be a nonempty set in a group G. Define

$$N(S) = \{x \in G \mid xSx^{-1} = S\}$$

and call $N(S)$ the <u>normalizer</u> of S. Show that $C(S) \subset N(S)$ and that $N(S)$ is a subgroup of G. Show that if S is a subgroup of G, then $S \subset N(S)$ and S is a normal subgroup of $N(S)$.

6. Show that if $\{H_\alpha \mid \alpha \in A\}$ is any collection of subgroups of G, then their intersection is also a subgroup of G. If W is any subset of G, by the <u>subgroup generated by</u> W we mean the intersection of all subgroups of G which contain W. Show that this is the smallest subgroup of G which contains W.

7. Consider two specific elements of $G = GL(n,2)$

$$A = \begin{pmatrix} 1 & 0 \\ 1 & 1 \end{pmatrix} \qquad B = \begin{pmatrix} 1 & 1 \\ 0 & 1 \end{pmatrix} .$$

Let H be the subgroup of G generated by A and K be the subgroup of G generated by B. Prove that $H = \{\ldots, A^{-2}, A^{-1}, I, A, A^2, \ldots\}$, and similarly for K.

8. Continuing with exercise 7, show that the product set

$$HK = \{hk \mid h \in H,\ k \in K\}$$

is not a subgroup of G. (Show that $ABAB$ is not of the form A^rB^s.)

9. We say that a subgroup K of G <u>normalizes</u> a subgroup H of G if for each $k \in K$ we have $kHk^{-1} = H$. Prove that if K normalizes H, then KH is a subgroup of G.

10. We can define an injective map

$$\phi : C \to M_2(R)$$

as follows: represent $\alpha \in C$ as $\alpha = \rho e^{i\theta}$ with $\rho \geq 0$ and set

$$\phi(\alpha) = \sqrt{\rho}\begin{pmatrix} \cos\theta & \sin\theta \\ -\sin\theta & \cos\theta \end{pmatrix} .$$

Show that $\phi(\alpha\beta) = \phi(\alpha)\phi(\beta)$ but that $\phi(\alpha+\beta)$ need not equal $\phi(\alpha)+\phi(\beta)$.

11. Let G be the multiplicative group of complex numbers of unit length. We say that $\alpha \in G$ is a <u>primitive n^{th} root of unity</u> if $\alpha^n = 1$, but none of $\alpha, \alpha^2, \ldots, \alpha^{n-1}$ are equal to one. Show that an isomorphism of G onto itself must send primitive n^{th} roots of unity to primitive n^{th} roots of unity for each n. For each n, how many primitive n^{th} roots of unity are there in G?

12. Let $\alpha = (a_1, a_2, a_3)$ and $\beta = (b_1, b_2, b_3)$ be two elements in R^3. Take the two "purely imaginary" quaternions

$$\alpha' = a_1 i + a_2 j + a_3 k$$

$$\beta' = b_1 i + b_2 j + b_3 k .$$

Show that if α' and β' are multiplied as quaternions, then

$$\alpha'\beta' - \text{real part } (\alpha'\beta')$$

is just the usual cross product of vectors in R^3.

Chapter 2
Orthogonal Groups

A. Inner products

We have a consistent notion of conjugation for $\mathbb{R} \subset \mathbb{C} \subset \mathbb{H}$.
Namely,

$$\text{for } x \in \mathbb{R}, \quad \bar{x} = x .$$

$$\text{For } \alpha = x + iy \in \mathbb{C}, \quad \bar{\alpha} = x - iy .$$

$$\text{For } q = x + iy + jz + kw \in \mathbb{H}, \quad \bar{q} = x - iy - jz - kw .$$

We clearly have $\bar{\bar{\alpha}} = \alpha$ in all cases and

$$\overline{(\alpha + \beta)} = \bar{\alpha} + \bar{\beta} .$$

It is an exercise to prove that

$$\overline{\alpha\beta} = \bar{\beta}\bar{\alpha} .$$

Of course for $\mathbb{R}$ or $\mathbb{C}$ this is the same as

$$\overline{\alpha\beta} = \bar{\alpha}\bar{\beta} .$$

Let $k \in \{\mathbb{R}, \mathbb{C}, \mathbb{H}\}$ and define an <u>inner product</u> $\langle \, , \, \rangle$ on k^n by

$$\langle x,y\rangle = x_1\bar{y}_1 + \ldots + x_n\bar{y}_n .$$

<u>Proposition</u> 1: $\langle\ ,\ \rangle$ <u>has the following properties</u>:

(i) $\langle x,y+z\rangle = \langle x,y\rangle + \langle x,z\rangle$

(ii) $\langle x+y,z\rangle = \langle x,z\rangle + \langle y,z\rangle$

(iii) $a\langle x,y\rangle = \langle ax,y\rangle$, $\langle x,ay\rangle = \langle x,y\rangle\bar{a}$

(iv) $\overline{\langle x,y\rangle} = \langle y,x\rangle$

(v) $\langle x,x\rangle$ <u>is always a real number</u> ≥ 0 <u>and</u>
$\langle x,x\rangle = 0 \Leftrightarrow x = (0,\ldots,0)$.

(vi) If $e_1,\ldots,e_n$ <u>is the standard basis for</u> k^n
$(e_i = (0,\ldots,0,1,0,\ldots,0))$, <u>then</u>

$$\langle e_i,e_j\rangle = \delta_{ij} = \begin{cases} 1 & \text{if } i = j \\ 0 & \text{if } i \neq j \end{cases} .$$

(vii) <u>The inner product is nondegenerate</u>; <u>i.e.</u>,

<u>If</u> $\langle x,y\rangle = 0$ <u>for all</u> y , <u>then</u> $x = (0,\ldots,0)$;
<u>If</u> $\langle x,y\rangle = 0$ <u>for all</u> x , <u>then</u> $y = (0,\ldots,0)$.

<u>Proof</u>: Exercise.

<u>Definition</u>: The <u>length</u> $|x|$ of $x \in k^n$ is

$$|x| = \sqrt{\langle x,x\rangle} .$$

Recall that if $A \in M_n(k)$, its <u>conjugate</u> $\bar{A}$ is obtained by replacing each a_{ij} by $\overline{a_{ij}}$; its <u>transpose</u> ${}^tA$ is obtained by replacing each a_{ij} by a_{ji} . These two operations commute so that the symbol

$$ {}^{t}\bar{A} $$

(the conjugate transpose of A) is unambiguous.

Recall that for $\mathbb{H}^n$ we must operate on the right (since we defined (scalar)(vector) on the left). So we do the same for $\mathbb{R}^n$ and $\mathbb{C}^n$. Thus we use row vectors.

<u>Proposition</u> 2: <u>For any</u> $x,y \in k^n$ <u>and</u> $A \in M_n(k)$ <u>we have</u>

$$ \langle xA, y \rangle = \langle x, y\,{}^{t}\bar{A} \rangle . $$

<u>Proof</u>: Let $A = (a_{ij})$.

$$ xA = (x_1a_{11} + \ldots + x_na_{n1}, \ldots, x_1a_{1n} + \ldots + x_na_{nn}) $$

$$ y\,{}^{t}\bar{A} = (y_1\bar{a}_{11} + \ldots + y_n\bar{a}_{1n}, \ldots, y_1\bar{a}_{n1} + \ldots + y_n\bar{a}_{nn}) $$

Thus the left hand side $\langle xA, y \rangle$ equals

$$ (x_1a_{11} + \ldots + x_na_{n1})\bar{y}_1 + \ldots + (x_1a_{1n} + \ldots + x_na_{nn})\bar{y}_n , $$

and the right hand side $\langle x, y\,{}^{t}\bar{A} \rangle$ equals

$$ x_1(a_{11}\bar{y}_1 + \ldots + a_{1n}\bar{y}_n) + \ldots + x_n(a_{n1}\bar{y}_1 + \ldots + a_{nn}\bar{y}_n) . $$

It is easy to see that these contain exactly the same terms.

B. <u>Orthogonal groups</u>

Again let $k \in \{\mathbb{R}, \mathbb{C}, \mathbb{H}\}$.

<u>Definition</u>:

$\mathfrak{G}(n,k) = \{A \in M_n(k) \mid \langle xA, yA\rangle = \langle x,y\rangle$ for all $x,y \in k^n\}$.

<u>Proposition 3</u>: $\mathfrak{G}(n,k)$ <u>is a group</u> .

<u>Proof</u>: If $A,B \in \mathfrak{G}(n,k)$, then

$$\langle xAB, yAB\rangle = \langle xA, yA\rangle = \langle x,y\rangle$$

so that

$$AB \in \mathfrak{G}(n,k) .$$

Clearly the identity matrix I is in $\mathfrak{G}(n,k)$.

If $A \in \mathfrak{G}(n,k)$ we have

$$\langle e_iA, e_jA\rangle = \langle e_i, e_j\rangle = \delta_{ij} = \begin{cases} 1 & \text{if } i = j \\ 0 & \text{if } i \neq j \end{cases} .$$

Now e_iA is just the ith row of A and we see that $\langle e_iA, e_jA\rangle$ is just the ij entry in the product

$$A\,{}^t\bar{A} .$$

Thus $A\,{}^t\bar{A} = I$. But then ${}^t\bar{A}A$ is also the identity since ${}^t\overline{({}^t\bar{A}A)} = {}^t(\bar{A}\,{}^tA) = A\,{}^t\bar{A} = I$. Thus ${}^t\bar{A} = A^{-1}$, a left hand and right inverse for A . (More generally, we saw in section C of chapter I that for matrices a left inverse was automatically a right inverse.) Finally,

$$\langle xA^{-1}, yA^{-1}\rangle = \langle xA^{-1}A, yA^{-1}A\rangle = \langle x,y\rangle ,$$

showing that $A^{-1} \in \mathfrak{G}(n,k)$. q.e.d.

Definition: For $k = \mathbb{R}$ we write $\mathfrak{O}(n,k)$ as $\mathfrak{O}(n)$ and call it the orthogonal group. For $k = \mathbb{C}$ we write it as $U(n)$ and call it the unitary group. For $k = \mathbb{H}$ we write it as $Sp(n)$ and call it the symplectic group.

Proposition 4: Let $A \in M_n(k)$. Then the following conditons are equivalent:

(i) $A \in \mathfrak{O}(n,k)$

(ii) $\langle e_i A, e_j A \rangle = \delta_{ij}$

(iii) A sends orthonormal bases to orthonormal bases

(iv) The rows of A form an orthonormal basis

(v) The columns of A form an orthonormal basis

(vi) ${}^t\bar{A} = A^{-1}$.

Proof: Exercise.

Proposition 5: Let $A \in M_n(\mathbb{R})$. Then $A \in \mathfrak{O}(n) \Leftrightarrow A$ preserves lengths.

Proof: A preserves lengths $\Leftrightarrow \langle xA, xA \rangle = \langle x,x \rangle$ for all $x \in \mathbb{R}^n$. So $\Rightarrow$ is trivial. Conversely, we have

$$\langle (x+y)A, (x+y)A \rangle = \langle x+y, x+y \rangle = \langle x,x \rangle + \langle x,y \rangle + \langle y,x \rangle + \langle y,y \rangle$$

$$= \langle xA, xA \rangle + \langle xA, yA \rangle + \langle yA, xA \rangle + \langle yA, yA \rangle \ .$$

This gives $\langle x,y \rangle + \langle y,x \rangle = \langle xA.yA \rangle + \langle yA, xA \rangle$ and since $\langle \ , \ \rangle$ over $\mathbb{R}$ is symmetric, this proves

$$\langle xA, yA\rangle = \langle x, y\rangle \text{ , i.e., } A \in \mathbb{O}(n) .$$

<u>Proposition</u> 5 bis: <u>Proposition</u> 5 <u>also holds for</u> $\mathbb{C}$ <u>and</u> $\mathbb{H}$.

<u>Proof</u>: Calculate $\langle (e_i+e_j)A, (e_i+e_j)A\rangle$ just as above to get

$$\langle e_iA, e_jA\rangle + \langle e_jA, e_iA\rangle = 0 .$$

Then consider $x = x_ie_i + x_je_j$ and calculate $\langle xA, xA\rangle$. We get $x_i\overline{x_j}\langle e_iA, e_jA\rangle + x_j\overline{x_i}\langle e_jA, e_iA\rangle = 0$ and thus

$$\langle e_iA, e_jA\rangle(x_i\overline{x}_j - x_j\overline{x}_i) = 0$$

and this forces $\langle e_iA, e_jA\rangle = 0$. q.e.d.

Let us look at $\mathbb{O}(n)$, $U(n)$ and $Sp(n)$ for small n. $\mathbb{O}(1)$ is the set of all real numbers of length one, so $\mathbb{O}(1) = \{1, -1\}$. $U(1)$ is just the set of all complex numbers of length one. This is the circle group S^1. $Sp(1)$ is the group of all quaternions of unit length. If we define

$$S^{k-1} = \{x \in \mathbb{R}^k \mid |x| = 1\}$$

to be the unit $(k-1)$-sphere we see that

$$\mathbb{O}(1) = S^0 , \quad U(1) = S^1 , \quad Sp(1) = S^3 .$$

It is an interesting fact that these are the <u>only</u> spheres which can be groups.

<u>Proposition</u> 6: <u>If</u> $k \in \{\mathbb{R}, \mathbb{C}\}$ <u>and</u> $A \in \mathbb{O}(n,k)$, <u>then</u>

$$(\det A)(\overline{\det A}) = 1 .$$

<u>Proof</u>: $A\,{}^t\bar{A} = I \Rightarrow (\det A)(\det {}^t\bar{A}) = 1$, and clearly

$$\det {}^t\bar{A} = \det \bar{A} = \overline{\det A} \ . \qquad \text{q.e.d.}$$

Thus if $A \in \mathfrak{O}(n)$ $(= \mathfrak{O}(n,\mathbb{R}))$, then $\det A \in \{1,-1\}$. We define

$$SO(n) = \{A \in \mathfrak{O}(n) \mid \det A = 1\}$$

and call this the <u>special</u> <u>orthogonal</u> group (also called the <u>rotation</u> group). Similarly, we define

$$SU(n) = \{A \in U(n) \mid \det A = 1\}$$

and call this the <u>special</u> <u>unitary</u> group.

An example of an element of $\mathfrak{O}(2) - SO(2)$ is $\begin{pmatrix} 1 & 0 \\ 0 & -1 \end{pmatrix}$. This sends $e_1 = (1,0)$ to e_1 and sends $e_2 = (0,1)$ to $-e_2$. It is just the reflection in the first axis, and has determinant equal to -1 .

C. <u>The isomorphism question</u>

At the end of Chapter I we showed that two groups which were defined quite differently were isomorphic. We have now defined several series of groups ($GL(n,k)$ for $n = 1,2,\ldots$ and $k \in \{\mathbb{R},\mathbb{C},\mathbb{H}\}$ and $\mathfrak{O}(n,k)$ for $n = 1,2,\ldots$ and $k \in \{\mathbb{R},\mathbb{C},\mathbb{H}\}$) and our major goal is to find out which of these are isomorphic. The basic idea will be to develop <u>invariants</u> of matrix groups (dimension, rank, etc.), i.e., two groups which are isomorphic must have the same invariants. This will make it possible to say that certain groups are <u>not isomorphic</u>. But when two differently defined groups are indeed isomorphic, an isomorphism may be hard to find. This is why we will work so hard to develop invariants--to reduce as much as possible the cases where we must look for isomorphisms. In this section we will give one isomorphism.

Suppose you suspect that $Sp(1)$ and $SU(2)$ are isomorphic. How would you try to find an isomorphism? $Sp(1)$ is the set of all quaternions of unit length and $SU(2)$ is the set of all complex 2×2 matrices A such that $A\,{}^t\bar{A} = I$ and $\det A = 1$. The operation in $Sp(1)$ is multiplication of quaternions, in $SU(2)$ it is matrix multiplication.

Proposition 7: The map $\phi : M_n(\mathbb{H}) \to M_{2n}(\mathbb{C})$ defined in §D of Chapter I induces an isomorphism

$$\phi : Sp(1) \to SU(2) .$$

Proof: We have seen that ϕ induces an injective homomorphism of $GL(n,\mathbb{H})$ into $GL(2n,\mathbb{C})$, so restriction of ϕ to $Sp(1)$ is still an injective homomorphism. So we just need to show that

(i) $A \in Sp(1) \Rightarrow \phi(A) \in SU(2)$ and

(ii) every $B \in SU(2)$ is some $\phi(A)$ with $A \in Sp(1)$.

If $A = a + ib + jc + kd$ then $\phi(A) = \begin{pmatrix} a+ib & -c-id \\ c-id & a-ib \end{pmatrix}$ so that

$$\phi(A)\,{}^t\overline{\phi(A)} = \begin{pmatrix} a+ib & -c-id \\ c-id & a-ib \end{pmatrix}\begin{pmatrix} a-ib & c+id \\ -c+id & a+ib \end{pmatrix} = \begin{pmatrix} 1 & 0 \\ 0 & 1 \end{pmatrix}$$

since $a^2 + b^2 + c^2 + d^2 = 1$. Also $\det \phi(A) = 1$ so $\phi(A) \in SU(2)$.

Let $B = \begin{pmatrix} \alpha & \beta \\ \gamma & \delta \end{pmatrix} \in SU(2)$. Using $\det B = 1$ and the fact that the rows are orthogonal unit vectors, we find that

$$\delta = \bar{\alpha} \quad \text{and} \quad \gamma = -\bar{\beta}.$$

So, if $\alpha = a + ib$ and $\beta = -c-id$, we may take $A = a+ib+jc+kd$ and have $\phi(A) = B$ (and $a^2 + b^2 + c^2 + d^2 = 1$).

D. Reflections in $\mathbb{R}^n$

Let u be a unit vector in $\mathbb{R}^n$ and let

$$u^{\perp} = \{x \in \mathbb{R}^n \mid \langle x,u\rangle = 0\}$$

be its orthogonal complement.

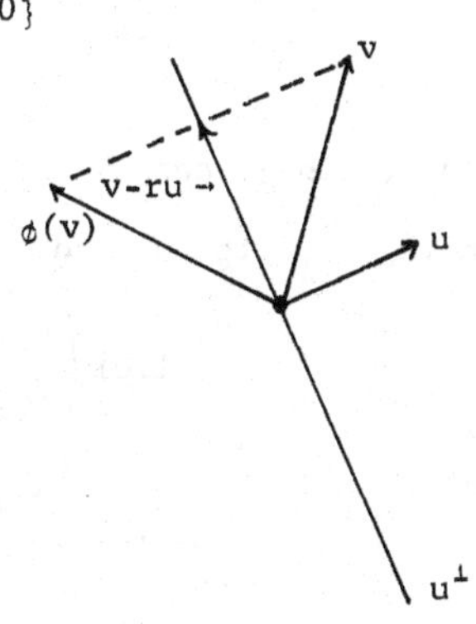

The projection of a vector v into $u^{\perp}$ is to be $v - ru$, where $r \in \mathbb{R}$ is to be chosen so that $v - ru$ is in $u^{\perp}$. So $0 = \langle v-ru,u\rangle = \langle v,u\rangle - r\langle u,u\rangle$ and thus

$$r = \langle v,u\rangle .$$

Clearly then the reflection of v in $u^{\perp}$ is to be

$$\phi(v) = v - 2ru = v - 2\langle v,u\rangle u .$$

Choose an orthonormal basis $u_1,\dots,u_n$ with $u_1 = u$. Then, using this basis, the reflection ϕ is given by the matrix $\begin{pmatrix} -1 & & & \bigcirc \\ & 1 & & \\ & & 1 & \\ \bigcirc & & & \ddots \, 1 \end{pmatrix}$. Let A be the linear map of $\mathbb{R}^n$ given by sending $e_1,\dots,e_n$ to $u_1,\dots,u_n$. By Proposition 4 A is orthogonal. So relative to our standard basis $e_1,\dots,e_n$ the reflection ϕ is given by

$$A^{-1}\begin{pmatrix} -1_1 & & \bigcirc \\ & \ddots & \\ \bigcirc & & 1 \end{pmatrix} A = {}^tA \begin{pmatrix} -1_1 & & \bigcirc \\ & \ddots & \\ \bigcirc & & 1 \end{pmatrix} A .$$

Conversely we see that such a matrix represents a reflection in the orthogonal complement of the vector e_1A .

In $\mathbb{R}^2$ let the unit vector u be written as

$$u = (\cos\alpha,\ \sin\alpha) .$$

Then $(-\sin\alpha,\ \cos\alpha)$ is a unit vector in $u^\perp$. The matrix A sending e_1 to u and e_2 to $(-\sin\alpha,\ \cos\alpha)$ must satisfy

$$(1,0)\begin{pmatrix} a_{11} & a_{12} \\ a_{21} & a_{22} \end{pmatrix} = (\cos\alpha,\ \sin\alpha)$$

$$(0,1)\begin{pmatrix} a_{11} & a_{12} \\ a_{21} & a_{22} \end{pmatrix} = (-\sin\alpha,\ \cos\alpha)$$

so

$$A = \begin{pmatrix} \cos\alpha & \sin\alpha \\ -\sin\alpha & \cos\alpha \end{pmatrix} .$$

Thus the matrix giving reflection in $u^\perp$ is

$$\Phi = \begin{pmatrix} \cos\alpha & -\sin\alpha \\ \sin\alpha & \cos\alpha \end{pmatrix}\begin{pmatrix} -1 & 0 \\ 0 & 1 \end{pmatrix}\begin{pmatrix} \cos\alpha & \sin\alpha \\ -\sin\alpha & \cos\alpha \end{pmatrix}$$

$$\Phi = \begin{pmatrix} -\cos 2\alpha & -\sin 2\alpha \\ -\sin 2\alpha & \cos 2\alpha \end{pmatrix} .$$

The matrix A is easily seen to be a rotation of $\mathbb{R}^2$ through an angle α .

Later in this course we will prove that $\mathcal{O}(n)$ is generated by reflections--that is, any element of $\mathcal{O}(n)$ may be obtained by a finite sequence of reflections.

E. <u>Exercises</u>

1. Prove Proposition 1.

2. Prove Proposition 4.

3. Let A be any element of $\mathcal{O}(n)$ with $\det A = -1$. Show that

$$\mathcal{O}(n) - SO(n) = \{BA \mid B \in SO(n)\} .$$

4. Show that any element of $SO(2)$ can be written as

$$\begin{pmatrix} \cos\theta & \sin\theta \\ -\sin\theta & \cos\theta \end{pmatrix} .$$

5. If $A \in U(n)$ and $\lambda \in \mathbb{C}$ has length one, show that $\lambda A \in U(n)$.

6. Let L_1 and L_2 be lines through the origin in $\mathbb{R}^2$. Show that reflection in L_1 followed by reflection in L_2 equals a rotation through twice the angle between L_1 and L_2 .

7. A matrix $A \in M_n(\mathbb{R})$ is said to be <u>idempotent</u> if $AA = A$. Show that the image of $\mathbb{R}^n$ under A is precisely the fixed-point set of A . Such a map is called a <u>projection</u> of $\mathbb{R}^n$ onto its image.

What can you say about the determinant of a idempotent matrix ? What is the image of $\mathbb{R}^2$ under $A = \begin{pmatrix} \frac{1}{2} & \frac{1}{2} \\ \frac{1}{2} & \frac{1}{2} \end{pmatrix}$.

8. A matrix A is <u>nilpotent</u> if some power of it is the zero matrix. For example $A = \begin{pmatrix} 0 & a & b \\ 0 & 0 & c \\ 0 & 0 & 0 \end{pmatrix}$ has its third power zero. Prove that a nilpotent matrix is singular. Prove that any $A = (a_{ij})$ with

$$a_{ij} = 0 \quad \text{whenever} \quad i \geqslant j$$

is nilpotent. Find two nilpotent matrices A and B whose product AB is not nilpotent.

9. Let U be the set of all matrices $A = (a_{ij})$ with all diagonal elements equal to one and

$$a_{ij} = 0 \quad \text{whenever} \quad i > j \ .$$

Prove that U is a group under matrix multiplication (the group of <u>unipotent</u> matrices in $M_n(\mathbb{R})$) .

Chapter 3
Homomorphisms

A. Curves in a vector space

We are going to define our first invariant of a matrix group, its dimension. Matrix groups whose dimensions are different can't be isomorphic. The dimension of a matrix group is going to be the dimension of its space of tangent vectors (a vector space), so we first define these.

Let V be a finite-dimensional real vector space. By a curve γ in V we mean a continuous function $\gamma:(a,b) \to V$ where (a,b) is an open interval in $\mathbb{R}$.

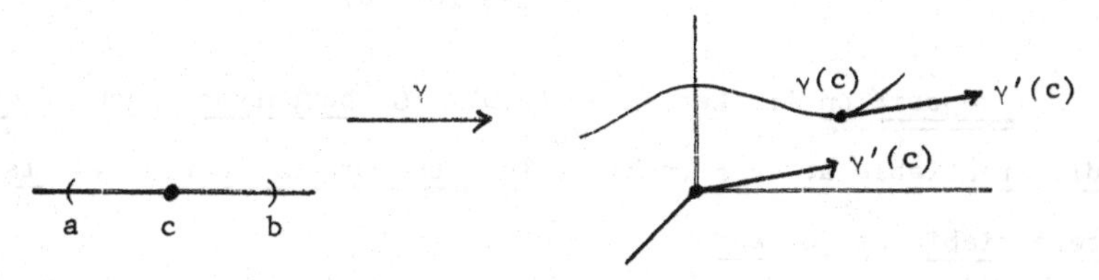

For $c \in (a,b)$ we say γ is differentiable at c if

$$\lim_{h\to 0} \frac{\gamma(c+h) - \gamma(c)}{h}$$

exists. When this limit exists, it is a vector in V. We denote it by $\gamma'(c)$ and call it the tangent vector to γ at $\gamma(c)$.

It is a standard result from calculus that if we choose a basis for V and thus represent γ as $(\gamma_1, \ldots, \gamma_n)$ (γ_i's being real valued), then $\gamma'(c)$ exists if and only if each $\gamma_i'(c)$ exists and

$$\gamma'(c) = (\gamma_1'(c), \ldots, \gamma_n'(c)) .$$

Now $M_n(\mathbb{R})$, $M_n(\mathbb{C})$, $M_n(\mathbb{H})$ can all be considered to be real vector spaces (of dimensions n^2, $2n^2$ and $4n^2$). If G is a matrix group in $M_n(k)$ then a <u>curve in</u> G is a curve in $M_n(k)$ with all values $\gamma(u)$ for $u \in (a,b)$ lying in G.

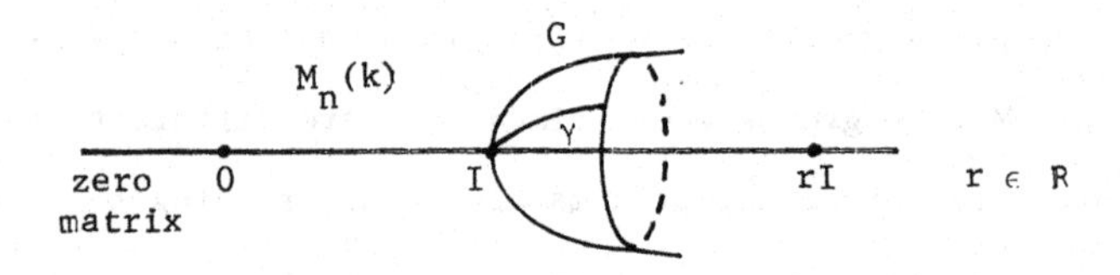

Suppose we have curves $\gamma, \sigma : (a,b) \to G$. Then we can define a new curve, the <u>product curve</u>, by

$$(\gamma\sigma)(u) = \gamma(u)\sigma(u) .$$

<u>Proposition</u> 1: <u>Let</u> $\gamma, \sigma : (a,b) \to G$ <u>be curves, both of which are differentiable at</u> $c \in (a,b)$. <u>Then the product curve</u> $\gamma\sigma$ <u>is differentiable at</u> c <u>and</u>

$$(\gamma\sigma)'(c) = \gamma(c)\sigma'(c) + \gamma'(c)\sigma(c) .$$

<u>Proof</u>: Let $\gamma(u) = (\gamma_{ij}(u))$, $\sigma(u) = (\sigma_{ij}(u))$. Then

$$(\gamma\sigma)(u) = (\sum_k \gamma_{ik}(u)\sigma_{kj}(u)) ,$$

so that

$$(\gamma\sigma)'(u) = (\sum_k \{\gamma'_{ik}(u)\sigma_{kj}(u) + \gamma_{ik}(u)\sigma'_{kj}(n)\})$$

$$= \gamma'(u)\sigma(u) + \gamma(u)\sigma'(u) .$$

Proposition 2: Let G be a matrix group in $M_n(k)$. Let T be the set of all tangent vectors $\gamma'(0)$ to curves $\gamma : (a,b) \to G$, $\gamma(0) = I$ $(0 \in (a,b))$. Then T is a subspace of $M_n(k)$.

Proof: If $\gamma'(0)$ and $\sigma'(0)$ are in T, then $(\gamma\sigma)(0) = \gamma(0)\sigma(0) = II = I$ and

$$(\gamma\sigma)'(0) = \gamma'(0)\sigma(0) + \gamma(0)\sigma'(0) = \gamma'(0) + \sigma'(0) .$$

Thus T is closed under vector addition.

T is also closed under scalar multiplication, for if $\gamma'(0) \in T$ and $r \in R$, let

$$\sigma(u) = \gamma(ru) .$$

Then $\sigma(0) = \gamma(0) = I$, σ is differentiable and $\sigma'(0) = r\gamma'(0)$.

Since $M_n(k)$ is a finite dimensional vector space, so is T.

Definition: If G is a matrix group, its dimension is the dimension of the vector space T (of tangent vectors to G at I).

Example 1: $U(1)$ has dimension 1.

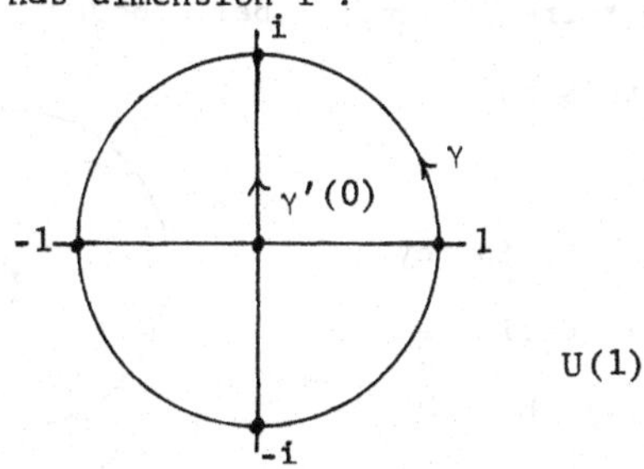

Example 2: dim Sp(1) = 3 .

Let $\gamma : (a,b) \to Sp(1)$ be a smooth curve with $\gamma(0) = 1$. Then $\gamma'(0)$ will be an element of $\mathbb{H} = R^4$. We first show $\gamma'(0)$ is in the span of i,j,k ; i.e. it is a quaternion with zero real part. Let

$$\gamma(t) = x(t) + i\, y(t) + j\, z(t) + k\, w(t)$$

with $x(0) = 1$ and $y(0) = 0$, $z(0) = 0$, $w(0) = 0$. We note that $x(0)$ is a maximum for the function x so that $\gamma'(0)$ $= 0 + i\, y'(0) + j\, z'(0) + k\, w'(0)$, as asserted.

Conversely, let $q = i\mu + j\nu + k\lambda$ be any quaternion with zero real part. We claim that there exists a smooth curve γ in Sp(1) such that $\gamma'(0) = q$. Indeed,

$$\gamma(t) = \sqrt{1 - \sin^2 \mu t - \sin^2 \nu t - \sin^2 \lambda t} + i \sin \mu t + j \sin \nu t + k \sin kt$$

can be readily verified to be such a curve (which is defined on some interval $[0,\epsilon)$, i.e. for t small).

Example 3: dim $GL(n,\mathbb{R}) = n^2$.

The determinant function $\det : M_n(\mathbb{R}) \to \mathbb{R}$ is continuous and $\det(I) = 1$. So there is some ϵ-ball about I in $M_n(\mathbb{R})$ such that for each A in this ball $\det A \neq 0$; i.e., $A \in GL(n,\mathbb{R})$. If v is any vector in $M_n(\mathbb{R})$ define a curve σ in $M_n(\mathbb{R})$ by

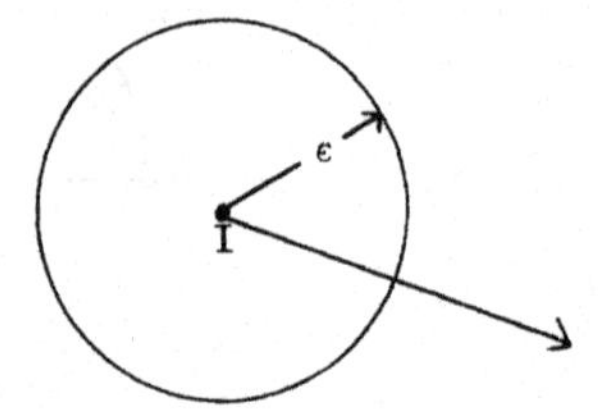

$$\sigma(t) = tv + I .$$

Then $\sigma(0) = I$ and $\sigma'(0) = v$ and for small t, $\sigma(t)$ is in $GL(n,\mathbb{R})$. Hence the tangent space T is all of $M_n(\mathbb{R})$ which has dimension n^2.

A similar argument shows that $\dim GL(n,\mathbb{C}) = 2n^2$.

We will now get upper bounds for the dimensions of $O(n)$, $U(n)$ and $Sp(n)$ after a few preliminaries.

<u>Definition</u>: $A \in M_n(\mathbb{R})$ is said to be <u>skew-symmetric</u> if

$$A + {}^tA = 0 ;$$

i.e. if $a_{ij} = -a_{ji}$ for each i,j. In particular, the diagonal terms must all be zero.

<u>Proposition</u> 3: <u>Let</u> $so(n)$ <u>denote the set of all skew-symmetric matrices in</u> $M_n(\mathbb{R})$. <u>Then</u> $so(n)$ <u>is a linear subspace of</u> $M_n(\mathbb{R})$, <u>and its dimension is</u> $\frac{n(n-1)}{2}$.

<u>Proof</u>: <u>The zero matrix is in</u> $so(n)$, <u>and if</u> A,B <u>belong to</u> $so(n)$, then

$$(A + B) + {}^t(A + B) = A + {}^tA + B + {}^tB = 0 ,$$

so that $so(n)$ is closed under vector addition. It is also closed under scalar multiplication, for if $A \in so(n)$ and $r \in \mathbb{R}$, then ${}^t(rA) = r\,{}^tA$ so that $rA + {}^t(rA) = r(A + {}^tA) = 0$.

To check the dimension of $so(n)$ we get a basis. Let E_{ij} denote the matrix whose entries are all zero except the ij entry, which is 1, and the ji entry, which is -1. If we define these

E_{ij} only for $i < j$. it is easy to see that they form a basis for $so(n)$, and it is easy to count that there are

$$(n-1) + (n-2) + \ldots + 1 = \frac{n(n-1)}{2} \quad \text{of them .}$$

<u>Definition</u>: A matrix $B \in M_n(\mathbb{C})$ is <u>skew-Hermitian</u> if

$$B + {}^t\bar{B} = 0 .$$

Thus if $b_{jk} = c + id$, then $\bar{b}_{kj} = -b_{jk} = -c - id$ and $b_{kj} = -c + id$. In particular if $j = k$ we have $c + id = -c + id$, so that the diagonal terms of a skew-Hermitian matrix are purely imaginary.

Let $su(n)$ be the set of skew-Hermitian matrices in $M_n(\mathbb{C})$. By the observation just made we see that $su(n)$ is not a vector space over $\mathbb{C}$.

<u>Proposition</u> 4: $su(n) \subset M_n(\mathbb{C})$ <u>is a real vector space of dimension</u>

$$n + 2\,\frac{n(n-1)}{2} = n^2 .$$

<u>Proof</u>: Exercise.

We make a similar definition for matrices in $M_n(\mathbb{H})$, and call $C \in M_n(\mathbb{H})$ <u>skew-symplectic</u> if

$$C + {}^t\bar{C} = 0 .$$

In the exercises one shows that the set $sp(n)$ of such matrices is a real vector space of dimension

$$3n + 4\,\frac{n(n-1)}{2} = n(2n+1) .$$

<u>Proposition</u> 5: <u>If</u> β <u>is a curve through the identity</u> $(\beta(0) = I)$

<u>in</u> $O(n)$ <u>then</u> $\beta'(0)$ <u>is skew-symmetric</u>

<u>in</u> $U(n)$ <u>then</u> $\beta'(0)$ <u>is skew-Hermitian</u>

<u>in</u> $Sp(n)$ <u>then</u> $\beta'(0)$ <u>is skew-symplectic</u> .

<u>Proof</u>: In each case we have that the product curve is constant

$$\beta(u)\,{}^t\bar{\beta}(u) = I \; .$$

Thus its derivative is zero, and the result follows from Proposition 1.

<u>Corollary</u>:

$$\text{Dim } O(n) \le \frac{n(n-1)}{2}$$

$$\text{Dim } U(n) \le n^2$$

$$\text{Dim } Sp(n) \le n(2n+1) \; .$$

Later we will show that these are equalities.

B. <u>Smooth homomorphisms</u>

Let $\phi : G \to H$ be a homomorphism of matrix groups. Since G and H are in vector spaces, it is clear what it means for ϕ to be continuous. From now on <u>homomorphism always means continuous homomorphism</u>. This being so, a curve

$$\rho : (a,b) \to G$$

gives a curve $\phi\circ\rho : (a,b) \to H$ by $(\phi\circ\rho)(u) = \phi(\rho(u))$ in H .

<u>Definition</u>: A homomorphism $\phi : G \to H$ of matrix groups is <u>smooth</u>

if for every differentiable curve ρ in G, $\phi\circ\rho$ is differentiable.

<u>Definition</u>: Let $\phi : G \to H$ be a smooth homomorphism of matrix groups. If $\gamma'(0)$ is a tangent vector to G at I we define a tangent vector $d\phi(\gamma'(0))$ to H at I by

$$d\phi(\gamma'(0)) = (\phi\circ\gamma)'(0) .$$

The resulting map $d\phi : T_G \to T_H$ is called the <u>differential</u> of ϕ.

<u>Proposition</u> 6: <u>$d\phi : T_G \to T_H$ is a linear map</u>.

<u>Proof</u>: Let $\rho'(0)$ and $\sigma'(0)$ be in T_G. For $a, b \in \mathbb{R}$ we define curves ρ_a, σ_b by $\rho_a(u) = \rho(au)$ and $\sigma_b(u) = \sigma(bu)$. Then $\rho_a'(0) = a\,\rho'(0)$ and $\sigma_b'(0) = b\,\sigma'(0)$.

Now, by definition,

$$d\phi[(\rho_a\sigma_b)'(0)] = [\phi\circ(\rho_a\sigma_b)]'(0).$$

Since ϕ is a homomorphism

$$\phi\circ(\rho_a\sigma_b) = (\phi\circ\rho_a)(\phi\circ\sigma_b).$$

Thus

$$\begin{aligned}(\phi\circ(\rho_a\sigma_b))'(0) &= (\phi\circ\rho_a)'(0)(\phi\circ\sigma_b)(0) \\ &\quad + (\phi\circ\rho_a)(0)(\phi\circ\sigma_b)'(0) \\ &= a\,d\,\phi(\rho'(0)) + b\,d\phi(\sigma'(0)).\end{aligned}$$

q.e.d.

<u>Proposition</u> 7: <u>If</u> $G \xrightarrow{\phi} H \xrightarrow{\psi} K$ <u>are smooth homomorphisms, then so is</u> $\psi\circ\phi$ <u>and</u>

$$d(\psi\circ\phi) = d\psi\circ d\phi .$$

<u>Proof</u>: The first part is obvious. For the second, let $\gamma'(0)$ be a tangent vector of G. Then

$$d(\psi\circ\phi)(\gamma'(0)) = (\psi\circ\phi\circ\gamma)'(0) = d\psi(\phi\circ\gamma)'(0) = d\psi\circ d\phi(\gamma'(0)) .$$

<u>Corollary</u>: <u>If</u> $\phi : G \to H$ <u>is a smooth isomorphism, then</u> $d\phi : T_G \to T_H$ <u>is a linear isomorphism and</u> $\dim G = \dim H$.

<u>Proof</u>: $\phi^{-1}\circ\phi$ is the identity, so $d\phi^{-1}\, d\phi : T_G \to T_G$ is the identity. Thus $d\phi$ is injective and $d\phi^{-1}$ is surjective. $\phi\,\phi^{-1}$ is the identity, so $d\phi\, d\phi^{-1} : T_H \to T_H$ is the identity. Thus $d\phi^{-1}$ is injective and $d\phi$ is surjective. q.e.d.

C. Exercises

1. Let $\gamma : (-1,1) \to M_3(\mathbb{R})$ be given by

$$\gamma(t) = \begin{pmatrix} \cos t & \sin t & 0 \\ -\sin t & \cos t & 0 \\ 0 & 0 & 1 \end{pmatrix} .$$

Show that γ is a curve in $SO(3)$ and find $\gamma'(0)$. Show that $(\gamma^2)'(0) = 2\gamma'(0)$.

2. Let $\sigma : (-1,1) \to M_3(\mathbb{R})$ be given by

$$\sigma(t) = \begin{pmatrix} 1 & 0 & 0 \\ 0 & \cos t & \sin t \\ 0 & -\sin t & \cos t \end{pmatrix} .$$

Calculate $\sigma'(0)$. Write the matrix for $\gamma(t)\sigma(t)$ and verify that $(\gamma\sigma)'(0) = \gamma'(0) + \sigma'(0)$.

3. Let $\rho : (-1,1) \to M_3(\mathbb{C})$ be given by

$$\rho(t) = \begin{pmatrix} e^{i\pi t} & 0 & 0 \\ 0 & e^{i\frac{\pi t}{2}} & 0 \\ 0 & 0 & e^{i\frac{\pi t}{2}} \end{pmatrix} .$$

Show that ρ is a curve in $U(3)$. Calculate $\rho'(0)$.

4. Let $\alpha : (-1,1) \to \mathbb{H}$ be defined by

$$\alpha(t) = (\cos t)j + (\sin t)k .$$

Show that α is in $Sp(1)$ and calculate $\alpha'(t)$.

5. Let H be a subgroup of a matrix group G. Show that T_H is a linear subspace of T_G so that $\dim H \leq \dim G$.

6. Show that the set $sp(n)$ of $n \times n$ skew-symplectic matrices is a real vector space and calculate its dimension.

7. Let T be the set of upper triangular matrices in $M_n(\mathbb{R})$. That is, $A = (a_{ij}) \in T$ if and only if $a_{ij} = 0$ whenever $i > j$. Show that T is a linear subspace of $M_n(\mathbb{R})$ and calculate its dimension. Show that T is a subalgebra of $M_n(\mathbb{R})$ (i.e. show that T is closed under multiplication). Show that $A \in T$ is nonsingular (i.e. is a unit) if and only if each $a_{ii} \neq 0$. (Note that the group U defined in Exercise 9 of Chapter 2 is a subgroup of the group of units in the algebra T.)

Chapter 4
Exponential and Logarithm

A. Exponential of a matrix

Given a matrix group G we have defined a vector space T -- the tangent space to G at I. In this chapter we develop maps to send T to G and G to T and study their properties. We will work with real matrices -- developments for $\mathbb{C}$ and $\mathbb{H}$ are quite analogous. (We need these maps to determine dimensions of some of our matrix groups.)

Definition: Let A be a real $n \times n$ matrix and set

$$e^A = I + A + \frac{A^2}{2!} + \frac{A^3}{3!} + \ldots$$

where A^2 means the matrix product AA, etc. We say that this sequence converges if each of the n^2 real-number sequences

$$(I)_{ij} + (A)_{ij} + (\frac{A^2}{2!})_{ij} + (\frac{A^3}{3!})_{ij} + \ldots$$

converges.

Proposition 1: *For any real $n \times n$ matrix A, the sequence*

$$I + A + \frac{A^2}{2!} + \ldots$$

converges.

Proof: Let m be the largest $|a_{ij}|$ in A. Then:

The biggest element in the first term is 1.

The biggest element in the second term is m.

The biggest element in the third term is $\leq \frac{nm^2}{2!}$.

The biggest element in the fourth term is $\leq \frac{n^2m^3}{3!}$, etc.

Any ij sequence is dominated by $1, m, \frac{nm^2}{2!}, \frac{n^2m^3}{3!}, \ldots,$

$\frac{n^{k-2}m^{k-1}}{(k-1)!}, \ldots$.

Applying the ratio test to this maximal sequence gives

$$\frac{n^{k-1}m^k}{k!} \frac{(k-1)!}{n^{k-2}m^{k-1}} = \frac{nm}{k} .$$

Since n and m are fixed, the ratio goes to 0 as $k \to \infty$, proving (absolute) convergence.

This exponential behaves somewhat like the familiar e^x $(x \in R)$. For if 0 is the zero matrix, we have

$$e^0 = I .$$

Also:

Proposition 2: If the matrices A and B commute, then

$$e^{A+B} = e^A e^B .$$

Proof: We will just indicate a proof by looking at the first few terms.

$$e^{A+B} = I + A + B + \frac{A^2}{2} + AB + \frac{B^2}{2} + \frac{A^3}{6} + \frac{A^2B}{2} + \frac{AB^2}{2} + \frac{B^3}{6} + \dots$$

$$e^Ae^B = (I + A + \frac{A^2}{2} + \frac{A^3}{6} + \dots)(I + B + \frac{B^2}{2} + \frac{B^3}{6} + \dots)$$

$$= I + A + B + \frac{A^2}{2} + AB + \frac{B^2}{2} + \frac{A^3}{6} + \frac{A^2B}{2} + \frac{AB^2}{2} + \frac{B^3}{6} + \dots .$$

Corollary: For any real $n \times n$ matrix A, e^A is nonsingular.

Proof: A and $-A$ commute, so $I = e^0 = e^{A-A} = e^Ae^{-A}$ and thus $1 = (\det e^A)(\det e^{-A})$ and $\det e^A \neq 0$.

From this corollary we see that the map $\exp : M_n(\mathbb{R}) \to M_n(\mathbb{R})$, $\exp(A) = e^A$, actually maps $M_n(\mathbb{R})$ into $GL(n, \mathbb{R})$.

Proposition 3: If A is a real skew-symmetric matrix, then e^A is orthogonal.

Proof: We have $I = e^0 = e^{A+{}^tA} = e^Ae^{{}^tA} = (e^A)^t(e^A)$, proving that e^A is orthogonal.

So, if $so(n) \subset M_n(\mathbb{R})$ is the subspace of skew-symmetric matrices, we see that

$$\exp : so(n) \to \mathbb{O}(n) .$$

It is important to note two things which Proposition 3 does not say: (i) it does not say that every orthogonal matrix is some e^A with A skew-symmetric (i.e. it does not say $\exp : so(n) \to \mathbb{O}(n)$ is surjective), and (ii) it does not say that e^A orthogonal implies A is skew-symmetric. It is instructive to examine the case $n = 2$ in some detail.

The general 2×2 real skew-symmetric matrix is of the form

$$\alpha = \begin{pmatrix} 0 & x \\ -x & 0 \end{pmatrix}, \quad x \in R .$$

To calculate e^{α}, we calculate the powers of α. $\alpha^2 = \begin{pmatrix} -x^2 & 0 \\ 0 & -x^2 \end{pmatrix}$, $\alpha^3 = \begin{pmatrix} 0 & -x^3 \\ x^3 & 0 \end{pmatrix}$, $\alpha^4 = \begin{pmatrix} x^4 & 0 \\ 0 & x^4 \end{pmatrix}$, $\alpha^5 = \begin{pmatrix} 0 & x^5 \\ -x^5 & 0 \end{pmatrix}$, etc. Then

$$e^{\alpha} = \begin{pmatrix} 1 & 0 \\ 0 & 1 \end{pmatrix} + \begin{pmatrix} 0 & x \\ -x & 0 \end{pmatrix} + \frac{1}{2!}\begin{pmatrix} -x^2 & 0 \\ 0 & -x^2 \end{pmatrix} + \frac{1}{3!}\begin{pmatrix} 0 & -x^3 \\ x^3 & 0 \end{pmatrix}$$
$$+ \frac{1}{4!}\begin{pmatrix} x^4 & 0 \\ 0 & x^4 \end{pmatrix} + \frac{1}{5!}\begin{pmatrix} 0 & x^5 \\ -x^5 & 0 \end{pmatrix} + \dots .$$

From the 1,1 position we get

$$1 - \frac{x^2}{2!} + \frac{x^4}{4!} - \frac{x^6}{6!} + \dots = \cos x , \text{ etc.}$$

We find that

$$e^{\alpha} = \begin{pmatrix} \cos x & \sin x \\ -\sin x & \cos x \end{pmatrix}$$

which is a plane rotation of x radians. Thus for any real 2×2 skew-symmetric matrix α we have

$$\det e^{\alpha} = 1 , \text{ i.e., } e^{\alpha} \in S\mathbb{O}(2) .$$

Thus, for example, the reflection $\begin{pmatrix} 1 & 0 \\ 0 & -1 \end{pmatrix} \in \mathbb{O}(2)$ could never be obtained this way.

Note also that $e^{\alpha} = I$ does not imply that α is the zero matrix ($\alpha = \begin{pmatrix} 0 & 2\pi \\ -2\pi & 0 \end{pmatrix}$ has $e^{\alpha} = I$).

We will see later that these results also hold for larger n.

We conclude this section with a simple observation which is sometimes quite useful in computations.

Proposition 4: *If* A,B *are* $n \times n$ *matrices over* $k \in \{R, C, \mathbb{H}\}$ *and* B *is nonsingular, then*

$$e^{BAB^{-1}} = Be^{A}B^{-1} .$$

Proof: $(BAB^{-1})^n = (BAB^{-1})(BAB^{-1})\ldots(BAB^{-1}) = BA^nB^{-1}$, and $B(C+D)B^{-1} = BCB^{-1} + BDB^{-1}$; these and the definition of the exponential of a matrix yield the result.

B. Logarithm

Just as e^x is defined for all $x \in R$ and $\log x$ is defined only for $x > 0$, the logarithm of a matrix will be defined only for matrices near to the identity matrix I .

Let X be a real $n \times n$ matrix and set

$$\log X = (X-I) - \frac{(X-I)^2}{2} + \frac{(X-I)^3}{3} - \frac{(X-I)^4}{4} + \ldots .$$

Proposition 5: For X near I this series converges.

Proof: Let $Y = X - I$ and $Y = (y_{ij})$, and suppose each $|y_{ij}| < \varepsilon$. Then

$$|(Y)_{ij}| \le \varepsilon , \quad |(\frac{Y^2}{2})_{ij}| \le \frac{n\varepsilon^2}{2} ,$$

$$|(\frac{Y^3}{3})_{ij}| \le \frac{n^2\varepsilon^3}{3} , \quad |(\frac{Y^k}{k})_{ij}| \le \frac{n^{k-1}\varepsilon^k}{k} .$$

The ratio test gives

$$\frac{n^k\varepsilon^{k+1}}{k+1} \frac{k}{n^{k-1}\varepsilon^k} = \frac{k}{k+1} n\varepsilon \to n\varepsilon .$$

So the series converges for any X such that each entry of $X - I$ is $< \frac{1}{n}$ in magnitude.

Proposition 6: In $M_n(R)$ let U be a neighborhood of I on *which* log *is defined and let* V *be a neighborhood of* 0 *such that* $\exp(V) \subset U$. *Then*

(i) *for* $X \in U$, $e^{\log X} = X$

(ii) *for* $A \in V$, $\log e^A = A$.

Proof: We do (ii) first. $A \in V \Rightarrow e^A \in U \Rightarrow \log e^A$ is defined (i.e., the series converges). $e^A - I = A + \frac{A^2}{2!} + \frac{A^3}{3!} + \ldots$. So

$$\log e^A = (A + \frac{A^2}{2!} + \ldots) - \frac{1}{2}(A + \frac{A^2}{2!} + \ldots)^2 + \frac{1}{3}(A + \frac{A^2}{2!} + \ldots)^3 + \ldots$$

$$= A + [\frac{A^2}{2!} - \frac{A^2}{2}] + [\frac{A^3}{6} - \frac{A^3}{2} + \frac{A^3}{3}] + \ldots = A .$$

(i) is similar. $\log X = (X-I) - \frac{(X-I)^2}{2} + \frac{(X-I)^3}{3} - \ldots$.

$$e^{\log X} = \{I + (X-I) - \frac{(X-I)^2}{2} + \ldots\} + \frac{1}{2!}\{(X-I) - \frac{(X-I)^2}{2} + \ldots\}^2$$

$$+ \frac{1}{3!}\{(X-I) - \frac{(X-I)^2}{2} + \ldots\}^3 + \ldots$$

$$= X - \frac{(X-I)^2}{2} + \frac{(X-I)^2}{2} + \{\frac{(X-I)^3}{3} - \frac{(X-I)^3}{2} + \frac{(X-I)^3}{6}\} + \ldots$$

$$= X .$$

Proposition 7: *If* X *and* Y *are near* I *and* $\log X$ *and* $\log Y$ *commute, then*

$$\log (XY) = \log X + \log Y \quad .$$

So if X *is near* I *and orthogonal,* $\log X$ *is skew-symmetric.*

Proof: $e^{\log XY} = XY = e^{\log X} e^{\log Y} = e^{\log X + \log Y}$, and e is one-to-one near 0 .

Now let X be orthogonal. Then X and ${}^tX$ commute so that $\log X$ and $\log {}^tX$ commute. Also

$$I = X^t X$$

so $0 = \log X^t X = \log X + \log {}^t X = \log X + {}^t(\log X)$ showing $\log X$ is skew-symmetric.

We will be able to say more after we have done a little topology. The picture will be

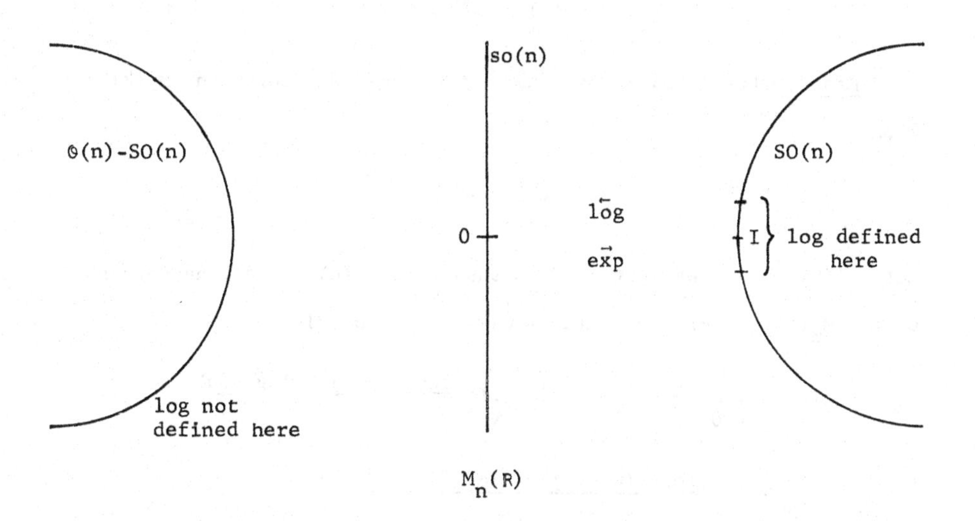

$M_n(\mathbb{R})$

C. <u>One-parameter subgroups</u>

<u>Definition</u>: A <u>one-parameter subgroup</u> γ in a matrix group G is a smooth homomorphism of the additive group $\mathbb{R}$ into G,

$$\gamma : \mathbb{R} \to G .$$

Note that it suffices to know γ on some open neighborhood U of 0 in $\mathbb{R}$. For $x \in \mathbb{R}$. some $\frac{1}{n} x \in U$ and $\gamma(x) = n(\gamma(\frac{1}{n}x))$.

Example: Let $k \in \{\mathbb{R}, \mathbb{C}, \mathbb{H}\}$ and $A \in M_n(k)$. Then

$$\gamma(u) = e^{uA} = I + uA + u^2 \frac{A^2}{2!} + \ldots$$

is a one-parameter subgroup of $GL(n,k)$ and $\gamma'(0) = A$.

Proposition 8: Let γ be a one-parameter subgroup of $GL(n,k)$. Then $\exists\ A \in M_n(k)$ such that

$$\gamma(u) = e^{uA} .$$

Proof: Let $\sigma(u) = \log \gamma(u)$. Then σ is a curve in $M_n(k)$ with

$$\gamma(u) = e^{\sigma(u)} .$$

Let $\sigma'(0) = A$. We just need to show that $\sigma(u)$ is a line through 0 in $M_n(k)$, for then $\sigma(u) = uA$. Hold u fixed.

$$\sigma'(u) = \lim_{v \to 0} \frac{\sigma(u+v) - \sigma(u)}{v} = \lim_{v \to 0} \frac{\log \gamma(u+v) - \log \gamma(u)}{v}$$

$$= \lim_{v \to 0} \frac{\log(\gamma(u)\gamma(v)) - \log \gamma(u)}{v} .$$

Now $u + v = v + u$ and γ is a one-parameter subgroup so that $\gamma(u)$ and $\gamma(v)$ commute. Thus

$$\log(\gamma(u)\gamma(v)) = \log \gamma(u) + \log \gamma(v) .$$

So

$$\sigma'(u) = \lim_{v \to 0} \frac{\log \gamma(v)}{v} = \sigma'(0) .$$

This proves that $\sigma'(u)$ is independent of u so $\sigma(u)$ is indeed a line through 0 in $M_n(k)$.

So any tangent vector to $GL(n,k)$ is the derivative at 0 of some one-parameter subgroup. We will see now that this is also true for the orthogonal groups $\mathfrak{O}(n,k)$.

Proposition 9: Let A be a tangent vector to $\mathfrak{O}(n,k)$. Then there exists a unique one-parameter subgroup γ in $\mathfrak{O}(n,k)$ such that

$$A = \gamma'(0) .$$

Proof: By definition $A = \rho'(0)$ where ρ is a curve in $\mathfrak{O}(n,k)$. Thus

$$\rho(u)\,{}^t\overline{\rho(u)} = I$$

so that

$$\rho'(0) + {}^t\overline{\rho'(0)} = 0 \text{ , i.e.,}$$

$$A + {}^t\bar{A} = 0 .$$

Now $\gamma(u) = e^{uA}$ is a one-parameter subgroup of $GL(n,k)$, but it lies in $\mathfrak{O}(n,k)$ because

$$\gamma(u)\ {}^t\overline{\gamma(u)} = e^{uA}e^{u\,{}^t\bar{A}} = e^{u(A+{}^t\bar{A})} = I .$$

This proves the proposition. So we have (for $GL(n,k)$ and $\mathfrak{O}(n,k)$) a one-to-one correspondence between tangent vectors and one-parameter subgroups.

Taking $k = \mathbb{R}$ we have that the tangent space to $\mathfrak{O}(n) = \mathfrak{O}(n,\mathbb{R})$ is $so(n)$, the vector space of all skew-symmetric $n \times n$ matrices. Thus $\dim \mathfrak{O}(n) = \dim so(n) = \frac{n(n-1)}{2}$.

Taking $k = \mathbb{C}$ we have that the tangent space to $U(n) = \mathfrak{O}(n,\mathbb{C})$

is $su(n)$, the vector space of all skew-Hermitian $n \times n$ complex matrices. Thus

$$\dim U(n) = \dim u(n) = n^2 .$$

Taking $k = \mathbb{H}$ we get

$$\dim Sp(n) = n(2n+1) .$$

What about the dimensions of $SO(n)$ and $SU(n)$? We will see in Chapter VI (Proposition 3) that the tangent space to $SO(n)$ is again $so(n)$, so the dimension of $SO(n)$ is also $\frac{n(n-1)}{2}$. But the dimension of $SU(n)$ is one less than the dimension of $U(n)$. The proof of this must also be deferred to a later chapter, but we will indicate here the result on which it is based.

<u>Definition</u>: The <u>trace</u> of a matrix $A = (a_{ij})$ is the sum of the diagonal terms;

$$Tr(A) = a_{11} + a_{22} + \dots + a_{nn} .$$

We clearly have

(i) $Tr(A+B) = Tr(A) + Tr(B)$, and $Tr(aA) = a\,Tr(A)$, (so Tr is linear).

Now suppose $A = (a_{ij})$ is real or complex. Then

(ii) $Tr(AB) = Tr(BA)$.

To prove (ii) we just write it out. The sum of the diagonal terms in AB is

$$(a_{11}b_{11}+\dots+a_{1n}b_{n1}) + (a_{21}b_{12}+\dots+a_{2n}b_{n2}) +\dots+ (a_{n1}b_{1n}+\dots+a_{nn}b_{nn})$$

and the sum of the diagonal terms in BA is

$$(b_{11}a_{11}+\ldots+b_{1n}a_{n1}) + (b_{21}a_{12}+\ldots+b_{2n}a_{n2}) +\ldots+ (b_{n1}a_{1n}+\ldots+b_{nn}a_{nn}) .$$

Since $\mathbb{R}$ and $\mathbb{C}$ are commutative one easily checks that these are equal.

Clearly

(iii) $Tr(I) = n$.

Also

(iv) If B is nonsingular, then

$$Tr(BAB^{-1}) = Tr(A) .$$

<u>Proof</u>: By (ii) , $Tr(B(AB^{-1})) = Tr((AB^{-1})B) = Tr(AI) = TrA$.

Now we come to the crucial relation.

<u>Theorem</u>: <u>If</u> A <u>is a real or complex matrix, then</u>

$$e^{Tr(A)} = \det (e^A) . \qquad (\dagger)$$

We will prove this later, but a few comments are in order here. First off, ($\dagger$) looks wrong because the left hand side depends only on the diagonal elements of A and it is not immediately clear that this is true for the right-hand side. The point is that det and e are also invariant under conjugation just as (iv) for Tr ; so if B is nonsingular

$$\det (e^{BAB^{-1}}) = \det (Be^AB^{-1}) = \det (e^A) .$$

We will prove ($\dagger$) once we have found how to put matrices in simpler forms by conjugation (in Chapter VIII).

Suppose we know (†). The linear map

$$\mathrm{Tr} : u(n) \to \mathbb{C}$$

actually maps into $i\mathbb{R} \subset \mathbb{C}$ since all diagonal terms in a skew-Hermitian matrix are purely imaginary. It is easy to see that $\mathrm{Tr}(u(n))$ is all of $i\mathbb{R}$. From the rank theorem in linear algebra we know that (all vector spaces being over $\mathbb{R}$ now)

$$\dim u(n) = \dim \mathrm{Tr}(u(n)) + \dim \mathrm{Tr}^{-1}(0) .$$

Thus the dimension of $\mathrm{Tr}^{-1}(0)$ is just one less than $\dim u(n)$ i.e. n^2-1. But $su(n) = \mathrm{Tr}^{-1}(0)$ is just the tangent space of $SU(n)$, since

$$\mathrm{Tr}(C) = 0 \Leftrightarrow 1 = e^{\mathrm{Tr}(C)} = \det e^C$$
$$\Leftrightarrow e^C \in SU(n) .$$

D. Lie algebras

It is easy to see that $so(n)$, $su(n)$ and $sp(n)$ are not closed under matrix multiplication. For example, if

$$\alpha = \begin{pmatrix} 0 & x \\ -x & 0 \end{pmatrix} \quad \text{then} \quad \alpha^2 = \begin{pmatrix} -x^2 & 0 \\ 0 & -x^2 \end{pmatrix}$$

which is not skew-symmetric.

Proposition 10: *For* $k \in \{\mathbb{R}, \mathbb{C}, \mathbb{H}\}$ *and* $A, B \in M_n(k)$ *we define*

$$[A,B] = AB - BA .$$

Then so(n), su(n) and sp(n) are closed under [,].

Proof: We need to show that

$$(AB - BA) + {}^t\overline{(AB - BA)} = 0 .$$

The left-hand side is

$$AB-BA+{}^t\bar{B}\,{}^t\bar{A}-{}^t\bar{A}\,{}^t\bar{B} = AB+(A\,{}^t\bar{B}-A\,{}^t\bar{B})-BA+(-B\,{}^t\bar{A}+B\,{}^t\bar{A})+{}^t\bar{B}\,{}^t\bar{A}-{}^t\bar{A}\,{}^t\bar{B}$$

$$= A(B+{}^t\bar{B})-(A+{}^t\bar{A})\,{}^t\bar{B}-B(A+{}^t\bar{A})+(B+{}^t\bar{B})\,{}^t\bar{A}$$

$$= 0 .$$

Thus so(n), su(n) and sp(n) become algebras (over $\mathbb{R}$) with this bracket multiplication. This product has some obvious properties.

(i) $[A,B] = -[B,A]$

(ii) $[A,B+C] = [A,B] + [A,C]$

$[A+B,C] = [A,C] + [B,C]$

(iii) For $r \in \mathbb{R}$, $r[A,B] = [rA,B] = [A,rB]$.

Finally, [,] has one nonobvious property.

(iv) $[A,[B,C]] + [B,[A,C]] + [C,[A,B]] = 0$.

Property (iv) is called the Jacobi identity and its proof is a routine verification.

Definition: A real vector space with a product satisfying (i)...(iv) is called a Lie algebra. (One could clearly consider complex Lie algebras, but we will have no occasion to do so.)

Let us consider low dimensional Lie algebras. For dim 1 the vector space is just $\mathbb{R}$ and if $x,y \in \mathbb{R}$ we have

$$[x,y] = x[1,y] = xy[1,1] = 0 \quad \text{(by (i))} .$$

So we have the trivial product (which obviously satisfies (i)...(iv)).

Consider $\mathbb{R}^2$ with basis e_1, e_2

We must have

$$[e_1,e_1] = 0 , \; [e_2,e_2] = 0 \quad \text{and} \quad [e_1,e_2] = -[e_2,e_1] .$$

Let $[e_1,e_2] = ae_1 + be_2$. Then, for example,

$$[e_1,[e_1,e_2]] = [e_1,(ae_1 + be_2)] = a[e_1,e_1] + b[e_1,e_2]$$

$$= b(ae_1 + be_2) .$$

By the Jacobi identity

$$[e_1,[e_1,e_2]] + [e_1,[e_2,e_1]] + [e_2,[e_1,e_1]] = 0 ,$$

so

$$b(ae_1 + be_2) + [e_1,(-ae_1 - be_2)] = 0$$

which is true with no conditions on a,b . If we take $a = 0 = b$ we get the trivial Lie algebra. For any other choice we get a nontrivial Lie algebra. In the exercises one shows that these nontrivial 2-dimensional Lie algebras are all "essentially the same."

We will not try to find out all nontrivial 3-dimensional Lie algebras, but will simply look at two which arise quite naturally.

$$so(3) = \left\{ \begin{pmatrix} 0 & a & b \\ -a & 0 & c \\ -b & -c & 0 \end{pmatrix} \middle| \; a,b,c \in \mathbb{R} \right\}$$

clearly has dimension three. Also using basis

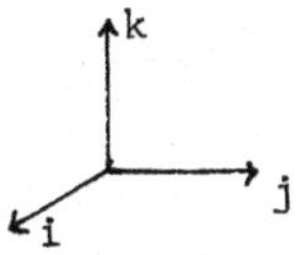

and defining

$$[i,j] = k$$
$$[j,k] = i$$
$$[k,i] = j$$

gives a 3-dimensional Lie algebra.

E. Exercises

1. Show that for A an $n \times n$ skew-symmetric matrix, every odd power of A is also skew-symmetric.

2. Let $B \in \mathfrak{G}(3) - SO(3)$. Show that the series for $\log B$ does not converge.

3. Prove the Jacobi identity for $[A,B] = AB - BA$.

4. Prove that any two nontrivial Lie algebras on $\mathbb{R}^2$ are isomorphic as Lie algebras.

5. Show that the two 3-dimensional Lie algebras defined above (§D) are isomorphic.

Chapter 5
SO(3) and Sp(1)

A. The homomorphism $\rho : S^3 \to SO(3)$

We have seen that $Sp(1)$, which is all quaternions of unit length, is just the unit 3-sphere in $\mathbb{R}^4$ $(= \mathbb{H})$. Also we have seen that $\dim SO(3) = \frac{3\cdot 2}{2} = 3$. So dimension won't distinguish S^3 from $SO(3)$, and, for all we know now, they might be isomorphic. In this section we define and study an "almost isomorphism" between them.

Proposition 1: *If $q \in S^3$, then the "left translation"*

$$L_q : \mathbb{H} \to \mathbb{H}$$

given by $L_q(q') = qq'$ is an orthogonal map of $\mathbb{R}^4$ to $\mathbb{R}^4$.

Proof: As vector spaces over $\mathbb{R}$, $\mathbb{H}$ and $\mathbb{R}^4$ are the same. So L_q is surely a *linear* map of $\mathbb{R}^4$, for if $a,b \in \mathbb{R}$ and $\alpha,\beta \in \mathbb{H}$ we have

$$L_q(a\alpha + b\beta) = q(a\alpha + b\beta) = aq\alpha + bq\beta = aL_q(\alpha) + bL_q(\beta) .$$

To see that L_q is orthogonal, it suffices to show that L_q preserves the perpendicularity (using $\langle\ ,\ \rangle$ for $\mathbb{R}^4$) of the four unit vectors $1,i,j,k$. For example, let $q = a + ib + jc + kd$ and calculate $\langle L_q(i),L_q(j)\rangle$ (using $\langle\ ,\ \rangle$ for $\mathbb{R}^4$). We get $ad + bc - bc - da = 0$. For $\langle L_q(1),L_q(i)\rangle$ we get

$$\langle a + ib + jc + kd,\ ai - b - kc + jd\rangle = -ab + ab + dc - dc = 0 \quad .$$

The computations for other pairs of basis vectors are similar.

<u>Definition of</u> ρ: For $q \in S^3$ and $\alpha \in \mathbb{H}$ we define

$$\rho(q)(\alpha) = q\alpha\bar{q} \ .$$

That is, we do a left translation by q and a right translation by $\bar{q}$. By Proposition 1 this is an orthogonal map of $\mathbb{R}^4$ to $\mathbb{R}^4$; i.e., $\rho(q) \in \mathfrak{O}(4)$.

Since real quaternions commute with all other quaternions, if x is a real quaternion

$$\rho(q)x = qx\bar{q} = xq\bar{q} = x \ .$$

Note also that $\rho(\bar{q})$ is the inverse of $\rho(q)$ in the group $\mathfrak{O}(4)$ since $\rho(q)\rho(q^{-1})(\alpha) = q(\bar{q}\alpha q)\bar{q} = \alpha$ and similarly for $\rho(\bar{q})\rho(q)$. Together these two observations imply that $\rho(q)$ maps the 3-space spanned by i,j,k to itself (Exercise #3). Thus $\rho(q)$ can be considered as an element of $\mathfrak{O}(3)$ (Exercise #4).

<u>Fact</u> (to be proved after Chapter VI): $\rho(q)$ is in $S\mathfrak{O}(3)$.

<u>Proposition</u> 2: $\rho : S^3 \to S\mathfrak{O}(3)$ <u>is a surjective homomorphism and</u>

$$\mathrm{Ker}(\rho) = \{1,-1\} \subset S^3 \ .$$

<u>Proof</u>: If $q_1,q_2 \in S^3$ and $\alpha \in \mathrm{Span}\langle i,j,k\rangle$, then

$$\rho(q_1q_2)(\alpha) = q_1q_2\alpha\overline{q_1q_2} = q_1(q_2\alpha\overline{q_2})\bar{q}_1 = \rho(q_1)\rho(q_2)(\alpha) .$$

Thus ρ is a homomorphism.

Clearly $\rho(1)$ and $\rho(-1)$ are the identity in SO(3) so that 1 and -1 are in Ker ρ . Conversely, suppose $\rho(q)$ is the identity with $q = a + ib + jc + kd$. Then $\rho(q)(i) = i$ gives $(a + ib + jc + kd)(i)(a - ib - jc - kd) = i$. And from this we get $a^2 + b^2 - c^2 - d^2 = 1$. But $a^2 + b^2 + c^2 + d^2 = 1$ and we conclude that $c = 0 = d$. From $\rho(q)j = j$ we get $b = 0$. Then $a^2 = 1$ so $a \in \{1,-1\}$.

Finally we need to show that ρ is surjective. This will be quite easy once we know some topology (Chapter VI)--otherwise it is an almost hopelessly complicated computation. Here we will just show that we can find a $q \in S^3$ such that $\rho(q)$ is the element of SO(3) which leaves k fixed, sends i to j and sends j to -i .

Let $q = a + ib + jc + kd$. We want
$(a + ib + jc + kd)(k)(a - ib - jc - kd) = k$, or
$(a + ib + jc + kd)(ka - jb + ic + d) = k$, so

$$ad - bc + bc - ad = 0 \quad \text{(automatically)} ,$$

$$ac + bd + ac + bd = 0 \quad \text{or} \quad 2(ac + bd) = 0 .$$

$$-ab + cd + dc - ab \quad \text{or} \quad 2(cd - ab) = 0 .$$

$$a^2 + d^2 - b^2 - c^2 = 1 .$$

Now

$$a^2 + d^2 + b^2 + c^2 = 1$$

so

$$2(b^2 + c^2) = 0 \quad \text{or} \quad b = 0 = c .$$

So the only condition on q such that $\rho(q)k = k$ is that $q = a + dk$ (with $a^2 + d^2 = 1$) .

Next we want $\rho(q)i = j$.

$$(a + kd)(i)(a - kd) = j$$

$$(a + kd)(ia + jd) = j$$

$$a^2 - d^2 = 0 , \ a = \pm d$$

$$ad + ad = 1 , \quad \text{if} \quad a = d , \ 2a^2 = 1 ,$$

and we can't have $a = -d$. Finally we insist that $\rho(q)j = -i$. So

$$(a + ka)(j)(a - ka) = -i$$

$$(a + ka)(ja - ia) = -1$$

$$-2a^2 = -1$$

$$a^2 - a^2 = 0 .$$

So $q = \frac{1}{\sqrt{2}} + k\frac{1}{\sqrt{2}}$ or $q = -\frac{1}{\sqrt{2}} - k\frac{1}{\sqrt{2}}$. Both will give the desired element of $SO(3)$. (This should be enough to convice us that we should not try the general proof of surjectivity at this stage.)

Note that this <u>does not prove</u> that S^3 and $SO(3)$ are not isomorphic. ρ is not an isomorphism, but one might exist. In the next section we give a fairly easy proof that $S^3 \not\cong SO(3)$.

B. Centers

In Exercise #4 of Chapter I the center C of a group G is defined as

$$C = \{x \in G \mid xy = yx \text{ for all } y \in G\},$$

and was shown to be an abelian and normal subgroup of G. We leave it as an exercise here to show that any isomorphisms of groups induces an isomorphism of their centers. We will show that $S^3 \not\cong SO(3)$ by showing that their centers are not isomorphic.

Proposition 3: The center of $S^3 = Sp(1)$ is $\{1,-1\}$, whereas the center of $SO(3)$ is $\{I\}$.

Proof: Since real quaternions commute with all quaternions, it is clear that $\{1,-1\} \subset \text{Center } S^3$. Conversely, suppose $q = a + ib + jc + kd \in S^3$ is in the center. Then $qi = iq$ gives

$$ai - b - ck + dj = ai - b + ck - dj$$

so that $c = 0 = d$. Then $qj = jq$ gives

$$(a + ib)j = j(a + ib)$$

and this implies $b = 0$. So $q = a$ and $a^2 = 1$. Thus Center $S^3 = \{1,-1\}$.

Suppose $A \in SO(3)$ is in the center. Since A commutes with all elements of $SO(3)$ it surely commutes with all elements of

$$T = \begin{pmatrix} \cos\theta & \sin\theta & 0 \\ -\sin\theta & \cos\theta & 0 \\ 0 & 0 & 1 \end{pmatrix}$$

since $T \subset SO(3)$. Consider the standard basis $e_1 = (1,0,0)$, $e_2 = (0,1,0)$, $e_3 = (0,0,1)$ for R^3 .

<u>Claim</u>: A leaves e_3 fixed (or sends it to $-e_3$) . Choose $B \in T$ which sends e_1 to e_2 and e_2 to $-e_1$ and (automatically) leaves e_3 fixed. Then set $Ae_3 = ae_1 + be_2 + ce_3$. Then $BAe_3 = ae_2 - be_1 + ce_3$, whereas $ABe_3 = Ae_3$; this implies $a = 0 = b$ and since A preserves length, we must have $c = 1$, or $c = -1$.

Thus A induces an orthogonal map of the e_1e_2 plane. Actually, it is a rotation because:

<u>Sublemma</u>: Any element of $\mathfrak{O}(2)$ which commutes with all rotations, is itself a rotation.

Let $\phi : R^2 \to R^2$ denote such an element of $\mathfrak{O}(2)$. For any rotation

$$t = \begin{pmatrix} \cos\theta & \sin\theta \\ -\sin\theta & \cos\theta \end{pmatrix}$$

we must have $\phi t = t\phi$. Let $\phi = \begin{pmatrix} \alpha & \beta \\ \gamma & \delta \end{pmatrix}$. We get

$$\alpha \cos\theta - \beta \sin\theta = \alpha \cos\theta + \gamma \sin\theta$$

$$\alpha \sin\theta + \beta \cos\theta = \beta \cos\theta + \delta \sin\theta ,$$

holding for all θ . So $\gamma = -\beta$ and $\alpha = \delta$. Thus

$$\phi = \begin{pmatrix} \alpha & \beta \\ -\beta & \alpha \end{pmatrix} \text{ and } \det\phi = \alpha^2 + \beta^2 .$$

Since this cannot equal -1 (and must be in $\{1,-1\}$), this proves the sublemma.

This also proves that $c = 1$ (not -1) (since $A \in SO(3)$) and we conclude that

$$A \in T .$$

We can now finish the proof.

$$A = \begin{pmatrix} \cos\theta & \sin\theta & 0 \\ -\sin\theta & \cos\theta & 0 \\ 0 & 0 & 1 \end{pmatrix}$$

and we let

$$R = \begin{pmatrix} 0 & 0 & 1 \\ 0 & 1 & 0 \\ -1 & 0 & 0 \end{pmatrix} \in SO(3) .$$

Since A must commute with R we get

$$AR = \begin{pmatrix} 0 & \sin\theta & \cos\theta \\ 0 & \cos\theta & -\sin\theta \\ -1 & 0 & 0 \end{pmatrix} = \begin{pmatrix} 0 & 0 & 1 \\ -\sin\theta & \cos\theta & 0 \\ -\cos\theta & -\sin\theta & 0 \end{pmatrix} = RA .$$

Thus we must have $\cos\theta = 1$ and $\sin\theta = 0$. Thus $A = I$ and Proposition 3 is proved.

We will calculate the centers of all of the groups $SO(n)$, $U(n)$, $SU(n)$, $Sp(n)$ in a later chapter, after we know about maximal tori. We conclude this chapter with a bit more abstract theory which we will need later.

C. Quotient groups

If H is a subgroup of G we define an equivalence relation $\sim$ on G by

$$x \sim y \text{ if } xy^{-1} \in H .$$

This relation is reflexive, $x \sim x$, since $xx^{-1} = e \in H$. It is symmetric, $x \sim y \Rightarrow y \sim x$, since $xy^{-1} \in H \Rightarrow (xy^{-1})^{-1} = yx^{-1} \in H$. It is transitive, $x \sim y$ and $y \sim z \Rightarrow x \sim z$, since $xy^{-1} \in H$ and $yz^{-1} \in H$ imply that $(xy^{-1})(yz^{-1}) = xz^{-1} \in H$. Thus $\sim$ divides G into equivalence classes.

Let $C(x)$ denote the class containing x. Then

$$C(x) = Hx = \{hx \mid h \in H\} .$$

Also $Hx = Hy \Leftrightarrow xy^{-1} \in H \Leftrightarrow y \in C(x) \Leftrightarrow x \in C(y)$. These equivalence classes are called right cosets of H.

Example: Let $G = S^3$ ($= Sp(1)$) and $H = \{1,-1\}$. Then $Hq = \{q,-q\} = H(-q)$ so each equivalence class contains exactly two points of S^3.

Example: In $G = U(3)$ let

$$H = \{\lambda I \mid \lambda \text{ a complex number of unit length}\} .$$

Then H is a circle subgroup of G and the right cosets are circles. Thus $U(3)$ can be divided into disjoint circles filling up $U(3)$.

Similarly, let $G = S^3 = Sp(1)$ and let H be the circle $\{a + ib \mid a^2 + b^2 = 1\}$. Thus S^3 can be divided up into circles.

One defines left cosets in a similar manner

$$xH = \{xh \mid h \in H\} .$$

Recall (Exercise #3, Chapter I) that a subgroup H is <u>normal</u> if $xHx^{-1} = H$ for all x in G.

<u>Observation</u>: A subgroup H of G is normal (in G) $\Leftrightarrow xH = Hx$ for every $x \in G$.

Let G/H denote the set whose elements are the rightcosets of H in G.

<u>Proposition</u> 4: <u>If H is a normal subgroup of G, then the operation on G/H defined by</u>

$$(Hx)(Hy) = H(xy)$$

<u>makes G/H into a group</u>.

<u>Proof</u>: We need H normal to show the operation on G/H is well defined. Suppose $Hx = Hz$ and $Hy = Hw$. We must show that $Hxy = Hzw$. Well, $xy(zw)^{-1} = xyw^{-1}z^{-1}$ and $yw^{-1} = h_1 \in H$. Also, $z^{-1} = x^{-1}h_2$. So

$$xy(zw)^{-1} = xh_1x^{-1}h_2$$

and, since H is normal, $xh_1x^{-1} = h_3 \in H$ so that

$$xy(zw)^{-1} = h_3h_2 \in H ,$$

and we have proved that the operation is well defined.

The rest is easy. $H = He \in G/H$ is the identity and Hx^{-1} is the inverse of Hx. (Associativity is inherited from G - $(Hx)(HyHz) = (HxHy)Hz$ since $x(yz) = (xy)z$).

Example: $G = Sp(1)$ and $H = \{1,-1\}$. H is the center of G and thus is a normal subgroup. Thus G/H is a group. We know it is $SO(3)$.

There is a natural map $\eta : G \to G/H$ given by $\eta(x) = Hx$. In the exercises it is shown that η is a surjective homomorphism with kernel H .

Let G be a group and $x,y \in G$. Then the element

$$xyx^{-1}y^{-1}$$

is called the commutator of x and y (because $(xyx^{-1}y^{-1})(yx) = xy$). Now the product of two commutators is not necessarily a commutator, but we set $[G,G] = \{\text{all finite products of commutators}\}$.

Proposition 5: $[G,G]$ is a normal subgroup of G and $\frac{G}{[G,G]}$ is an abelian group.

Proof: Closure and identity are clear and

$$(xyx^{-1}y^{-1})(yxy^{-1}x^{-1}) = e \ ,$$

showing $[G,G]$ is a subgroup. Let $z \in G$ and $xyx^{-1}y^{-1} \in [G,G]$. Then

$$z(xyx^{-1}y^{-1})z^{-1} = zxy(z^{-1}(xy)^{-1}(xy)z)x^{-1}((yz)^{-1}(yz))y^{-1}z^{-1}$$

$$= \{z(xy)z^{-1}(xy)^{-1}\}\{x(yz)x^{-1}(yz)^{-1}\}\{yzy^{-1}z^{-1}\} \in [G,G] \ .$$

This easily extends to products of commutators, so that $[G,G]$ is a normal subgroup.

Finally, $[G,G]\,x\,[G,G]y = [G,G]xy = [G,G]yx = [G,G]y[G,G]x$ since

$xy(yx)^{-1} = xyx^{-1}y^{-1} \in [G,G]$.

q.e.d.

In most instances we will encounter, if G is a matrix group and C is its center, then G/C will have trivial center. But this need not always be the case.

<u>Proposition</u> 6: <u>For</u> $x \in G$ <u>define</u>

$$\psi(x) : G \to G$$

<u>by</u> $\psi(x)(y) = xyx^{-1}y^{-1}$. <u>Then</u> G/C <u>has nontrivial center</u> $\Leftrightarrow \exists x \in G-C$ <u>such that</u>

$$\psi(x)(G) \subset C .$$

<u>Proof</u>: $\Leftarrow$

$x \notin C \Rightarrow Cx \neq C$ so Cx is not the identity in G/C . But for any $y \in G$ we have $xyx^{-1}y^{-1} \in C$ so that $CxCy = Cxy = Cyx = CyCx$ and $Cx \in$ center G/C .

$\Rightarrow$

Conversely, $Cx \neq C$ with Cx in the center implies

$CxCy = Cxy = Cyx = CyCx$ so that $xyx^{-1}y^{-1} \in C$ for all $y \in G$.

Once we have done a little topology (Chapter VI) we easily have:

<u>Corollary</u>: <u>If</u> G <u>is connected and</u> C <u>is discrete (in particular) if</u> C <u>is finite), then</u> G/C <u>has no center</u>.

D. Exercises

1. Let G be a group and $x \in G$. Show that left translation $L_x : G \to G$ by x $(L_x(g) = xg)$ is a one-to-one map of G onto G. Let R_x be right translation so that

$$R_{x^{-1}} \circ L_x(g) = xgx^{-1} .$$

Show that $R_{x^{-1}} \circ L_x$ is an isomorphism of G onto G.

2. Do one more step in the proof of Proposition 1 by showing $\langle L_q(i), L_q(k) \rangle = 0$.

3,4. These are listed by number in the text.

5. Show that $\rho(i)$, $\rho(j)$, $\rho(k)$ are all in $SO(3)$.

6. Show that the set T defined in the proof of Proposition 3 is an abelian subgroup of $SO(3)$.

7. Let $\phi : G \to K$ be a surjective homomorphism of groups and $H = \operatorname{Ker} \phi$. Then we have

$$\begin{array}{l} G \xrightarrow{\phi} K \\ \downarrow \eta \\ G/H \end{array} .$$

Show that $\phi \circ \eta^{-1}$ is well defined and gives an isomorphism of G/H onto K.

8. Show that (see Exercise #6) the abelian subgroup T of $SO(3)$ is not a normal subgroup.

9. Show that the subgroup $H = \{a + ib \mid a^2 + b^2 = 1\}$ of $Sp(1)$

is not a normal subgroup.

10. Show an isomorphism of groups induces an isomorphism of their centers.

Chapter 6
Topology

A. Introduction

Our matrix groups are all subsets of euclidean spaces, because they are all subsets of

$$M_n(\mathbb{R}) = \mathbb{R}^{n^2} \quad \text{or} \quad M_n(\mathbb{C}) = \mathbb{R}^{2n^2} \quad \text{or} \quad M_n(\mathbb{H}) = \mathbb{R}^{4n^2} .$$

There are certain _topological_ properties, notably connectedness and compactness, which some of our groups have and others do not. These properties are preserved by continuous maps and so are surely invariants under isomorphisms of groups. So a connected matrix group could not be isomorphic with a nonconnected matrix group, and a similar statement holds for compactness. We will define these properties and decide which of our groups have them. This will be done in sections B and C.

In section D we define and discuss the notion of a countable basis for open sets, a concept we will need in our study of maximal tori in matrix groups. Finally, in section E we define _manifold_ and show that all of our matrix groups are manifolds. Then we prove a theorem about manifolds which gives an easy proof that the homomorphism $\rho : Sp(1) \to SO(3)$ (defined in Chapter V) is surjective.

B. Continuity of functions, open sets, closed sets

Definition: A metric d on a set S is a way of assigning to each $x, y \in S$ a real number $d(x,y)$ (the distance from x to y) in such a way that:

(i) $d(x,y) \geq 0$ and $d(x,y) = 0 \Leftrightarrow x = y$.

(ii) $d(x,y) = d(y,x)$.

(iii) $d(x,y) + d(y,z) \geq d(x,z)$.

Condition (iii) is called the triangle inequality.

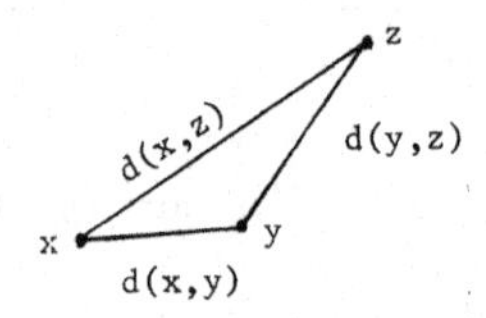

We will define such a metric d on $\mathbb{R}^n$ and then for any $S \subset \mathbb{R}^n$. d will also clearly be a metric on S. Recall that for $x, y \in \mathbb{R}^n$,

$$x = (x_1, \ldots, x_n) \qquad y = (y_1, \ldots, y_n)$$

we defined an inner product

$$\langle x,y \rangle = x_1 y_1 + \ldots + x_n y_n .$$

Set $d(x,y) = \sqrt{\langle x-y, x-y \rangle}$. (Thus we define $d(x,y)$ to be the length of the vector $x - y$.)

Proposition 1: This is a metric on $\mathbb{R}^n$.

Proof: Properties (i) and (ii) follow from

$$\langle x,x \rangle \geq 0 \text{ and } \langle x,x \rangle = 0 \Leftrightarrow x = 0$$

and symmetry of the inner product. To prove the triangle inequality we will prove the corresponding property of $\langle\ ,\ \rangle$ called the <u>Schwarz inequality</u>.

For any $x,y \in \mathbb{R}^n$ and $t \in \mathbb{R}$ we have

$$\langle x+ty, x+ty\rangle \geq 0 .$$

Using the bilinearity and symmetry of $\langle\ ,\ \rangle$ this gives

$$\langle x,x\rangle + 2\langle x,y\rangle t + \langle y,y\rangle t^2 \geq 0 .$$

This quadratic polynomial in t with real coefficients is always ≥ 0 and thus it cannot have two distinct real roots. (A quadratic polynomial can have only one minimum.) Thus the discriminant cannot be positive; i.e.,

$$(2\langle x,y\rangle)^2 - 4\langle y,y\rangle\langle x,x\rangle \leq 0 .$$

So

$$(*) \qquad \langle x,y\rangle^2 \leq \langle x,x\rangle\langle y,y\rangle .$$

The inequality (*) is the Schwarz inequality.

We apply (*) to the vectors $x - y$ and $y - z$ to get

$$(**) \qquad \langle x-y, y-z\rangle \leq \sqrt{\langle x-y, x-y\rangle}\,\sqrt{\langle y-z, y-z\rangle} .$$

If we square both sides of (iii), write it in terms of $\langle\ ,\ \rangle$ and use basic properties of $\langle\ ,\ \rangle$, we see that (iii) is equivalent to (**).

We use this metric d on $\mathbb{R}^n$ to define <u>open balls</u>. Let $x \in \mathbb{R}^n$ and $r > 0$ be a real number. Set

$$B(x,r) = \{y \in \mathbb{R}^n \mid d(x,y) < r\}$$

and call this the open ball with center x and radius r. Open balls in euclidean spaces allow us to give a fairly direct generalization of the notion of continuity of a function on $\mathbb{R}$ to functions defined on spaces of dimension greater than one.

Let A be a subset of $\mathbb{R}^n$ and

$$f : A \to \mathbb{R}^m$$

be a function defined on A and taking values in some euclidean space $\mathbb{R}^m$.

Definition: To say f is continuous at a point $a \in A$ means: Given any open ball $B(f(a),\varepsilon)$ in $\mathbb{R}^m$, there exists an open ball $B(a,\delta)$ in $\mathbb{R}^n$ such that any point $x \in A \cap B(a,\delta)$ satisfies

$$f(x) \in B(f(a),\varepsilon) .$$

Another way of saying this is: Given $\varepsilon > 0$ there exists $\delta > 0$ such that if $x \in A$ satisfies $d(a,x) < \delta$, then $f(x)$ satisfies $d(f(a),f(x)) < \varepsilon$.

Both ways are just precise ways of saying that f is continuous if it sends "nearby points" of A to "nearby points" in $\mathbb{R}^m$.

It is important to notice that the continuity of f depends on the domain A of definition. For example define

$$f : \mathbb{R} \to \mathbb{R}$$

by

$$f(x) = \begin{cases} 0 & \text{if } x < 0 \\ 1 & \text{if } x \geq 0 \end{cases} .$$

Then f is not continuous at 0. But suppose we take $A \subset \mathbb{R}$ to be all $x \geq 0$ and restrict f to A

$$f : A \to R .$$

Then this restricted f is continuous at 0.

A function $f : A \to R^m$ $(A \subset R^n)$ is said to be <u>continuous</u> if it is continuous at each $a \in A$.

<u>Example</u>: If $A \subset R^n$ is a finite set then any $f : A \to R^m$ is continuous.

<u>Proof</u>: For any $a \in A$ let $b_1, \ldots, b_k$ be all of the other points of A. Let

$$\delta_i = d(a, b_i) \ , \quad i = i, \ldots, k$$

and let δ be the smallest of these. Then for any $\varepsilon > 0$ any element of A in $B(a, \delta)$ goes into $B(f(a), \varepsilon)$ (because a is the only such element of A).

<u>Proposition</u> 2: <u>If</u> $A \subset R^n$ <u>and</u>

$$A \xrightarrow{f} R^m \ , \ f(A) \xrightarrow{g} R^p$$

<u>are continuous, then</u> $g \circ f$ <u>is continuous.</u>

<u>Proof</u>: Let $a \in A$ and $\varepsilon > 0$. Since g is continuous, there exists $\eta > 0$ such that every element of $f(A)$ in $B(f(a), \eta)$ is sent by g into $B(g(f(a)), \varepsilon)$. Since f is continuous there exists $\delta > 0$ such that every element of A in $B(a, \delta)$ is sent by f into $B(f(a), \eta)$ and is then sent by g into $B(g(f(a)), \varepsilon)$.

q.e.d.

Some exercises on continuity are given at the end of this chapter.

Definition: A set $U \subset R^n$ is an open set if each $x \in U$ lies in some $B(x,r) \subset U$ (where r will depend on x).

Clearly R^n is an open set. It is not quite so clear that the empty set ϕ is open; but since there is no $x \in \phi$, there is no requirement that some $B(x,r)$ be contained in ϕ.

Example: Any open ball $B(y,s)$ is an open set. Let $x \in B(y,s)$ (i.e., $d(x,y) < s$). We must find $r > 0$ such that $B(x,r) \subset B(y,s)$. Well, let $r = s - d(x,y)$. If $z \in B(x,r)$ then $d(z,x) < s - d(x,y)$

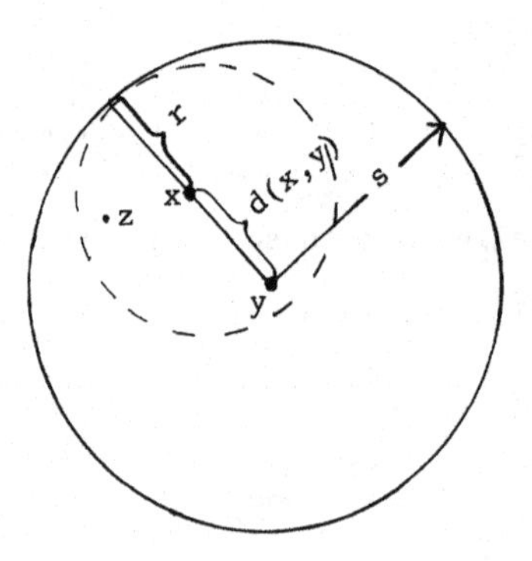

and thus

$$d(z,x) + d(z.y) < s .$$

Using the triangle inequality we have

$$d(z.y) \leq d(z,x) + d(x,y) < s$$

showing $z \in B(y.s)$.

Example: $(0,1) = \{x \in R \mid 0 < x < 1\}$ is an open set in R (because it is the open ball $B(\frac{1}{2}.\frac{1}{2})$). But

$(0,1] = \{x \in \mathbb{R} \mid 0 < x \leq 1\}$ is not open in $\mathbb{R}$ because $1 \in (0,1]$ but no $B(1,r)$ lies in $(0,1]$ since every such ball contains numbers greater than 1.

Definition: A subset $C \subset \mathbb{R}^n$ is defined to be <u>closed</u> if its complement $\mathbb{R}^n - C$ is open.

$(0,1] \subset \mathbb{R}$ is neither open nor closed. We have seen it is not open. Let $T = \mathbb{R} - (0,1]$. Then $0 \in T$ but no $B(0,r)$ can lie in T since each will contain points of $(0,1]$. Thus T is not open so that $(0,1]$ is not closed.

$[0,1] = \{x \in \mathbb{R} \mid 0 \leq x \leq 1\}$ is a closed set.

Example: Let $K \subset \mathbb{R}^n$ be a finite set. Then K is closed. For if $x \in \mathbb{R}^n - K$ there will be a minimum distance δ from x to points of K. Then

$$B(x,\delta) \subset \mathbb{R}^n - K,$$

proving K is closed.

C. <u>Connected sets, compact sets</u>

Definition: A set D in $\mathbb{R}^n$ is <u>connected</u> if: Given $x,y \in D$ there exists a continuous function

$$\gamma : [0,1] \to D$$

$$(\text{i.e., } \gamma : [0,1] \to \mathbb{R}^n \text{ with } \gamma([0,1]) \subset D)$$

with $\gamma(0) = x$ and $\gamma(1) = y$.

Such a function may be called a <u>path</u> from x to y in D.

<u>Examples</u>: $\mathbb{R}^n$ is connected because

$$\gamma(t) = (x + t(y-x))$$

is a path in $\mathbb{R}^n$ from x to y.

$D = \{x \in \mathbb{R} \mid x \neq 0\}$ is not connected. For example, no path from -1 to 1 in $\mathbb{R}$ can lie in D.

$D = \{(x_1,x_2) \in \mathbb{R}^2 \mid (x_1,x_2) \neq (0,0)\}$ is connected.

<u>Important example</u>: $\mathcal{O}(n) \subset \mathbb{R}^{n^2}$ is not connected. The matrices

$$A = \begin{pmatrix} -1 & & & \bigcirc \\ & 1 & & \\ & & \ddots & \\ \bigcirc & & & 1 \end{pmatrix} \quad \text{and} \quad I = \begin{pmatrix} 1 & & & \bigcirc \\ & 1 & & \\ & & \ddots & \\ \bigcirc & & & 1 \end{pmatrix}$$

are in $\mathcal{O}(n)$. If $\gamma : [0,1] \to \mathcal{O}(n)$ is a path from A to I, then the composite function

$$[0,1] \xrightarrow{\gamma} \mathcal{O}(n) \xrightarrow{\det} \mathbb{R}$$

would be continuous (Proposition 2). But it would be a path in $\{-1,1\} \subset \mathbb{R}$ from -1 to 1, contradicting the existence of γ.

Recall that $so(n) \subset M_n(\mathbb{R})$ consists of skew-symmetric matrices and that if $A \in so(n)$, then $\exp A \subset \mathcal{O}(n)$.

<u>Proposition</u> 3: exp <u>maps</u> so(n) <u>into</u> SO(n).

<u>Proof</u>: For $B \in so(n)$ the path

$$\gamma(t) = e^{tB}$$

is a path from $e^0 = I$ to e^B. As seen above, this implies that $\det e^B = +1$ so $e^B \in SO(n)$.

<u>Proposition</u> 4: <u>Let</u> $D \subset \mathbb{R}^n$ <u>be connected and</u>

$$f : D \to \mathbb{R}^m$$

<u>be continuous, then</u> $f(D)$ <u>is connected.</u>

<u>Proof</u>: Given $a,b \in f(D)$, choose $x,y \in D$ such that $f(x) = a$ and $f(y) = b$. Choose a path γ from x to y in D . Then $f \circ \gamma$ is a path from a to b in $f(D)$.

<u>Definition</u>: A subset W of $\mathbb{R}^n$ is <u>bounded</u> if W lies in some open ball. This is clearly equivalent to: W lies in some $B(0,r)$.

Now boundedness, unlike connectedness, is not preserved by continuous functions. For example, if $W = (0,1) \subset \mathbb{R}$ and $f : W \to \mathbb{R}$ is defined by $f(x) = \frac{1}{x}$, then W is bounded but $f(W)$ is not bounded.

Neither is the property of being closed preserved by continuous functions. For example, $\mathbb{R}$ is closed in $\mathbb{R}$ and $f : \mathbb{R} \to \mathbb{R}$ defined by $f(x) = e^x$ is continuous, but $f(\mathbb{R}) = \{y \in \mathbb{R} \mid y > 0\}$ is not closed.

However, when we put closed and bounded together, they are then both preserved.

<u>Definition</u>: $C \subset \mathbb{R}^n$ is <u>compact</u> if it is closed and bounded.

<u>Proposition</u> 5: <u>If</u> $C \subset \mathbb{R}^n$ <u>is compact and</u>

$$f : C \to \mathbb{R}^m$$

<u>is continuous, then</u> $f(C)$ <u>is compact.</u>

The proof is relegated to an appendix.

D. Subspace topology, countable bases

Sometimes we will have a subset W of R^n and will want to know which subsets of W we should call open sets in W.

Definition: If $U \subset W \subset R^n$ we say U is an open set in W if there exists an open set V in R^n such that

$$U = V \cap W .$$

For example, if $W = [0,1] \subset R$, then $U = (\frac{1}{2},1] = \{x \in R \mid \frac{1}{2} < x \leq 1\}$ is an open set in W, but not an open set in R. $U' = [\frac{1}{2},1]$ is not open in W.

Note that if W is an open set in R^n, then $U \subset W$ is open in W if and only if it is open in R^n.

For $W \subset R^n$ the collection of all open sets of W is the subspace topology of W.

Recall that $V \subset R^n$ is defined to be open if any $x \in V$ has some $B(x,r) \subset V$. This is equivalent to saying that $V \subset R^n$ is open if it is either the empty set or is a union of open balls. (Exercise.) Of course, not every open set is an open ball, but open balls suffice to give all open sets by taking unions (the "empty" union being allowed).

Definition: A collection $\mathcal{U} = \{V_\alpha\}$ of open sets in R^n is a basis for open sets if every open set in R^n is a union of some of the V_α's.

<u>Examples</u>: The set of all open squares in $\mathbb{R}^2$ is a basis for the open sets in $\mathbb{R}^2$.

The set of all open intervals $(a,b) \subset \mathbb{R}$ with a and b rational is a basis for the open sets in $\mathbb{R}$.

The set of all open balls in $\mathbb{R}^n$ is, of course, a basis for open sets. But so is

$$\{B(x,r) \mid x = (x_1,\ldots,x_n) \text{ with each } x_i \text{ rational, and } r \text{ is rational}\}$$

(See Proposition 7.)

For a subset W of $\mathbb{R}^n$ we know which are the open sets in W and we can give the same definition as above for the notion of a basis for the open sets in W. Indeed, it is clear that if $\mathcal{V} = \{V_\alpha\}$ is a basis for the open sets in $\mathbb{R}^n$, then $\{V_\alpha \cap W\}$ is a basis for the open sets of W.

We want to get bases for open sets which are "minimial" in the sense that they have no more sets than needed to do the job. The notion that comes up is <u>countability</u>.

<u>Definition</u>: A set S is <u>countable</u> if its elements can all be arranged in a finite or infinite sequence $s_1, s_2, s_3, \ldots$; that is, every element of S will be somewhere in the sequence.

<u>Examples</u>: The set $\mathbb{Q}$ of all positive rational numbers is countable; for example

$$1,\ 2,\ \frac{1}{2},\ \frac{3}{2},\ 3,\ \frac{1}{3},\ \frac{2}{3},\ \frac{4}{3},\ \frac{5}{3},\ \frac{7}{3},\ \frac{8}{3},\ 4,\ \frac{1}{4},\ \ldots$$

is a sequence containing all positive rationals. Similarly,

$1, -1, 2, -2, \frac{1}{2}, -\frac{1}{2}, \frac{3}{2}, -\frac{3}{2}, 3, -3, \ldots$ contains all of $\mathbb{Q}$.

The set $I = [0,1] = \{x \in \mathbb{R} \mid 0 \le x \le 1\}$ is not countable. We prove this contrapositively. Suppose

$$r_1, r_2, r_3, \ldots$$

is a list of all elements of I . It suffices to give an element of I which cannot be in the list. Express the r_i's as decimals

$$r_i = .x_{i_1} x_{i_2} x_{i_3} \ldots \quad .$$

Let $r = .y_1 y_2 y_3 \ldots$ where $y_j = 5$ if $x_{jj} \ne 5$ and $y_j = 1$ if $x_{jj} = 5$. Then

$$r \ne r_1 \text{ because } y_1 \ne x_{11}$$

$$r \ne r_2 \text{ because } y_2 \ne x_{22} \text{ , etc.}$$

But $r \in I$.

<u>Proposition</u> 6: <u>If A and B are countable sets, then so is their cartesian product</u> $A \times B$.

<u>Proof</u>: Let $A = \{a_1, a_2, a_3, \ldots\}$ and $B = \{b_1 b_2 . b_3, \ldots\}$. Then we can write

$$A \times B = \{(a_1,b_1),(a_1,b_2),(a_2,b_1),(a_3,b_1),(a_2,b_2), \ldots \}$$

because following the path shown below will include all of $A \times B$.

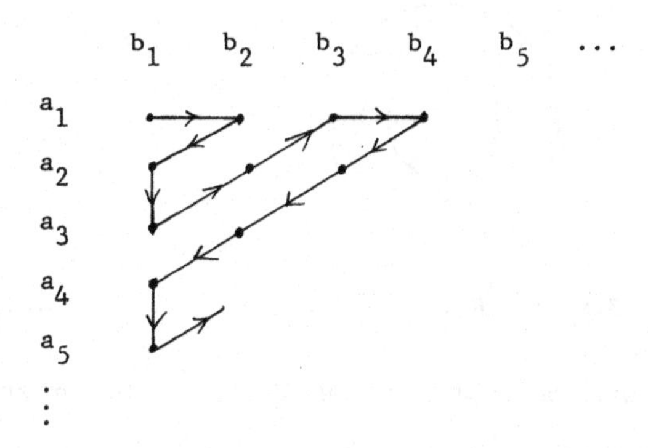

Proposition 7: R^n (and hence any $W \subset R^n$) has a countable basis for its open sets.

Proof: The set

$$C = \{B(x,r) \mid x = (x_1,\ldots,x_n) \text{ each } x_i \in \mathbb{Q} \text{ and } r \in \mathbb{Q}\}$$

can be put in 1-1 correspondence with (n+1)-tuples $(x_1,\ldots,x_n,r)$ of rational numbers (with $r > 0$). By Proposition 6 this is a countable set of balls.

Let V be any open set in R^n. To show that V is a union of elements of C it suffices to show that for $y \in V$ some $B(x,r) \in C$ contains y and lies in V. (For then V is the union of such $B(x,r)$ -- one for each $y \in V$.) Since V is open some

$$B(y,s) \subset V .$$

Choose x with all coordinates rational such that

$$d(x,y) < \frac{s}{3}$$

and let r be a rational number satisfying $\frac{s}{3} < r < \frac{s}{2}$.

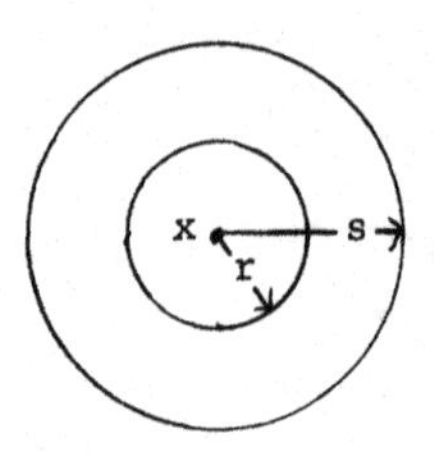

Then $y \in B(x,r)$ and $B(x,r) \subset B(y,s) \subset V$. q.e.d.

This proposition will be used in an essential way in our study of maximal tori in matrix groups.

E. Manifolds

Definition: By a space we mean some subset of some $\mathbb{R}^n$ with the subspace topology. A map

$$f : X \to Y$$

of spaces is a homeomorphism if it is one-to-one, f is continuous, and f^{-1} is continuous.

Example: $f(x) = e^{ix}$ is a one-to-one continuous map of $[0,2\pi) \subset \mathbb{R}$ to the unit circle $S^1 \subset \mathbb{R}^2$. But f is not a homeomorphism because f^{-1} is not continuous. (It "tears" the circle open.)

Example: Let G be a matrix group and $x \in G$. Then left translation L_x by x $(L_x(g) = xg)$ is a homeomorphism $L_x : G \to G$. (Exercise.)

A manifold is a space which "locally" looks like some $\mathbb{R}^n$.

Definition: A space X is an n-manifold if each $x \in X$ lies in some open set homeomorphic to some $B(o,r) \subset R^n$. An n-manifold is said to have dimension n.

Proposition 8: A matrix group of dimension n is an n-manifold.

Proof: The exponential map from the n-dimensional tangent space T to G is continuous. It is one-to-one on some neighborhorhood V of 0 in T because it has an inverse (log). Also this inverse is continuous. Take $B(o,r) \subset V$ and we have that the identity matrix I has the right kind of neighborhood. For any $x \in G$ we have

$$L_x \circ \exp : B(o,r) \to G$$

being a homeomorphism (composition of homeomorphisms is a homeomorphisim) onto a neighborhood of x. Thus G is an n-manifold.

Definition: A manifold is called closed if it is compact (= closed and bounded).

Proposition 9: $GL(n,k)$ is not closed, but $\mathfrak{G}(n,k)$ is closed.

Proof: Clearly $GL(n,k)$ is not bounded, because for every non-zero real number r, $rI \in GL(n,k)$. (This also shows that $GL(n,k)$ is not closed. Because $0 \in M_n(k) - GL(n,k)$ but every ball with center 0 will contain some $rI \in GL(n,k)$.)

If $A \in \mathfrak{G}(n,k)$ then the rows are unit vectors so that as a vector in $M_n(k)$ the length of A is $\leq n$. Thus $\mathfrak{G}(n,k)$ is a bounded set.

To see that $M_n(k) - \mathfrak{G}(n,k)$ is open, suppose $\tilde{B} \in M_n(k) - \mathfrak{G}(n,k)$. Then there exists $x,y \in k^n$ such that

$$\langle x\tilde{B}, y\tilde{B}\rangle \neq \langle x,y\rangle .$$

Since $\langle\ ,\ \rangle$ is continuous, there is some open ball $B(\tilde{B},s)$ in $M_n(k)$ such that for $B' \in B(\tilde{B},s)$ we have $\langle xB', yB'\rangle \neq \langle x,y\rangle$. Thus $B' \notin \mathfrak{G}(n,k)$.

q.e.d.

We finish this chapter with a result which will be of substantial use to us later on.

Proposition 10: Let N and M be closed n-manifolds with $N \subset M$. If M is connected, then $N = M$.

Proof: We want to show that $M - N$ is empty. If it isn't, choose $y \in M - N$ and $x \in N$. Since M is connected, there exists a path

$$\rho : [0,1] \to M$$

with $\rho(0) = x$ and $\rho(1) = y$. Then $\rho^{-1}(M - N)$ is an open set in $[0,1]$ (see Exercise 3) and it contains 1 but not 0. Let t_0 be the largest element of the closed set $I - \rho^{-1}(M - N)$. Then

(i) every $B(t_0,\varepsilon)$ contains points of $\rho^{-1}(M - N)$, but

(ii) since N is a manifold there is some open neighborhood U of $\rho(t_0)$ in N. By continuity of ρ some $B(t_0,\varepsilon)$ maps by ρ into N. This contradiction shows $N = M$.

Corollary: The map $\rho : Sp(1) \to SO(3)$ (see Chapter V) is surjective.

Proof: Since ρ is a homeomorphism on some neighborhood of each point, we see that the image $\rho(Sp(1))$ is a 3-manifold. Since ρ is continuous, this image is a closed 3-manifold (Proposition 5). It remains to prove that $SO(3)$ is connected (so that we may apply Proposition 10). It suffices to show that any $A \in SO(3)$ may be joined to the identity matrix I by a path in $SO(3)$.

We have $\det A = 1$ and that $\{Ae_1, Ae_2, Ae_3\}$ is an orthonormal basis for $\mathbb{R}^3$. Let B be a rotation sending e_1 to Ae_1 and leaving the direction perpendicular to the (e_1, Ae_1) plane fixed. (If $Ae_1 = e_1$ proceed directly to the next step. If e_1 and Ae_1 are antipodal on S^2, then we have two choices for B.) Clearly, there is a path ω from I to B in $SO(3)$. Now $Be_1 = Ae_1$ so Be_2 and Be_3 are an orthonormal basis for the plane perpendicular to Ae_1. Let C be a rotation of this plane sending Be_2 to Ae_2 and Be_3 to Ae_3. (If we can't do this, we would have $\det A = -1$.) There is a path σ from I to C in $SO(3)$. Since $A = BC$, we can multiply the paths ω and σ to get a path from I to A in $SO(3)$.

q.e.d.

<u>Note</u>: In Chapter VIII we will prove that $SO(n)$ is connected for all n.

F. Exercises

1. Show that the definition of continuity reduces to the usual $\varepsilon - \delta$ definition for $f : (a,b) \to \mathbb{R}$.

2. Suppose we have $A \subset \mathbb{R}^n$ and have functions

$$A \xrightarrow{f} \mathbb{R}^m \qquad f(A) \xrightarrow{g} \mathbb{R}^p .$$

We have seen that f and g continuous implies that $g \circ f$ is continuous. Give examples to show that:

f continuous and $g \circ f$ continuous $\not\Rightarrow g$ continuous .

g continuous and $g \circ f$ continuous $\not\Rightarrow f$ continuous .

3. Show that for $A \subset R^n$ and $f : A \to R^m$ then f is continuous $\Leftrightarrow$ for each open set U in R^m , $f^{-1}(U)$ is an open set in A .

4. Show that if A,B are connected sets in R^n and $A \cap B \neq \phi$, then $A \cup B$ is connected.

5. Let H be any connected subgroup of a matrix group G . Show that

$$S = \bigcup_{x \in G} xHx^{-1}$$

is connected.

6. Show that matrix multiplication is continuous (with one matrix fixed; i.e., $A \in M_n(k)$, $L_A : M_n(k) \to M_n(k)$ given by $L_A(B) = AB$ is continuous) .

7. Show that an arbitrary union of open sets is an open set.

8. Let $A \subset R^n$ and $x \in R^n$. We say that x is a limit point of A (x lp A) if every

$$B(x,r) \cap A$$

is an infinite set. Show that $C \subset R^n$ is a closed set $\Leftrightarrow$ (x lp $C \Rightarrow x \in C$) .

9. Let $D \subset \mathbb{R}^n$ be open and closed. Show that if D is not empty, then $D = \mathbb{R}^n$. (See the proof of Proposition 10.)

Chapter 7
Maximal Tori

A. Cartesian products of groups

If G and H are groups, we make $G \times H$ into a group by defining

$$(g,h)(g',h') = (gg',hh') .$$

This works and if G,H are abelian so is $G \times H$. (Exercises.)

Example: If G is a group of $n \times n$ matrices and H is a group of $m \times m$ matrices, we can represent elements of $G \times H$ as $(n + m) \times (n + m)$ matrices by

$$(g,h) = \begin{array}{c} n\{ \\ \\ \end{array} \left(\begin{array}{c|c} g & 0 \\ \hline \underbrace{0}_{n} & \underbrace{h}_{m} \end{array} \right) \}m .$$

Then matrix multiplication gives the operation described above on $G \times H$. (Exercise.) Let us look at an important special case of this.

Let $G = \{\begin{pmatrix} \cos\theta & \sin\theta \\ -\sin\theta & \cos\theta \end{pmatrix}\}$ and $H = \{\begin{pmatrix} \cos\phi & \sin\phi \\ -\sin\phi & \cos\phi \end{pmatrix}\}$. Then

$$G \times H = \left\{ \begin{pmatrix} \cos\theta & \sin\theta & 0 & 0 \\ -\sin\theta & \cos\theta & 0 & 0 \\ 0 & 0 & \cos\phi & \sin\phi \\ 0 & 0 & -\sin\phi & \cos\phi \end{pmatrix} \right\} .$$

Both G and H are circle groups, and the (abelian) group $G \times H = S^1 \times S^1$ is called a _2-torus_.

Definition: A _k-torus_ is the Cartesian product of k circle groups.

We have seen that a k-torus can be represented by a "block diagonal" $2k \times 2k$ real matrix. But it is easy to see that

$$T = \begin{pmatrix} e^{i\theta_1} & & & \\ & e^{i\theta_2} & & \bigcirc \\ & \bigcirc & \ddots & \\ & & & e^{i\theta_k} \end{pmatrix}$$

is a k-torus, so we can represent a k-torus as diagonal complex $k \times k$ matrices.

Proposition 1: If G is an abelian matrix group and γ, σ are one-parameter subgroups, then $\gamma\sigma$ is a one-parameter subgroup.

Proof:

$$\begin{aligned}(\gamma\sigma)(s+t) &= \gamma(s+t)\sigma(s+t) \\ &= \gamma(s)\gamma(t)\sigma(s)\sigma(t) \\ &= \gamma(s)\sigma(s)\gamma(t)\sigma(t) \\ &= (\gamma\sigma)(s)(\gamma\sigma)(t)\ .\end{aligned}$$

Corollary: If G is an abelian matrix group then $\exp : (TG)_e \to G$ is a homomorphism from the vector group of the tangent space $(TG)_e$ to G at e .

Proof: Let $\xi = \gamma'(0)$, $\eta = \sigma'(0)$ with γ, σ being one-parameter subgroups. Then from Chapter IV we know that $\exp(\xi) = \gamma(1)$ and

$\exp(\eta) = \sigma(1)$. We have that $\gamma\sigma$ is a one-parameter subgroup and $(\gamma\sigma)'(0) = \gamma'(0) + \sigma'(0) = \xi + \eta$ (Chapter III), so $\exp(\xi+\eta) = (\gamma\sigma)(1) = \gamma(1)\sigma(1) = \exp \xi \exp \eta$.

Now we know that exp is 1-1 on some neighborhood V of 0 in $(TG)_e$. So for G abelian we know that $\ker(\exp) = L$ is a <u>discrete</u> subgroup of the vector group $(TG)_e$ (i.e. some neighborhood of 0 contains no point of L except 0). Next we consider when $\exp : (TG)_e \to G$ is surjective.

<u>Proposition</u> 2: <u>Let G be a connected matrix group and let H be any subgroup of G containing an open neighborhood U of e . Then H = G .</u>

<u>Proof</u>: Since $U \subset H$ and H is a subgroup, we must have

$$U^2 = \{xy \mid x,y \in U\},\ U^3,\ U^4,\ \ldots \text{ all in } H .$$

Thus
$$W = U \cup U^2 \cup U^3 \cup \ldots \subset H .$$

Each U^k is open, so W is an open set. (Exercise, Chapter VI.) But W is also closed. For, let x be a limit point of W . Then xU is an open set containing x $(e \in U)$ and hence must contain some point of W . Thus

$$xu = u_1 \ldots u_m \text{ for some } u, u_1, u_2, \ldots, u_m \in U .$$

But then $x = u_1 \ldots u_m u^{-1} \in W$. In a connected space G only ϕ and G are both open and closed (Exercise, Chapter VI), $W \neq \phi$ so $W = G$ so $H = G$.

<u>Corollary</u>: If G is a connected abelian matrix group, then

$\exp : (TG)_e \to G$ is a surjective homomorphism with a discrete kernel.

Proof: exp is a homomorphism so $\exp((TG)_e)$ is a subgroup of G and it contains some neighborhood U of e. Thus $\exp(TG)_e = G$.

In the exercises it is proved that this implies that $\exp(TG)_e = T^k \times R^{n-k}$ for some k. So we have

Theorem 1: Any compact connected abelian matrix group G is a torus.

B. Maximal tori in groups

Definition: A subgroup H of a matrix group G is a torus if it is isomorphic with a k-torus for some k. It is a maximal torus if it is not contained in any larger torus subgroup of G.

Proposition 3:

$$T = \left\{ \begin{pmatrix} \cos\theta & \sin\theta & 0 \\ -\sin\theta & \cos\theta & 0 \\ 0 & 0 & 1 \end{pmatrix} \right\}$$

is a maximal torus in $SO(3)$.

Proof: Clearly T is isomorphic with $\{\begin{pmatrix} \cos\theta & \sin\theta \\ -\sin\theta & \cos\theta \end{pmatrix}\}$ which is a circle group and thus is a 1-torus.

Suppose there is a larger torus subgroup T' of $SO(3)$, i.e.,

$$T \subsetneq T' \subset SO(3).$$

Since T' would be abelian, we would have: $\phi \in SO(3)$ such that

$\phi \notin T$ but ϕ commutes with every element of T. So it suffices to prove that:

If $\phi \in SO(3)$ commutes with each $t \in T$, then $\phi \in T$. Refer back to our proof that Center $SO(3) = \{I\}$ in Chapter V, and you will see that we have already proved this fact.

Proposition 4:

$$T = \left\{ \begin{pmatrix} \cos\theta_1 & \sin\theta_1 & 0 & 0 \\ -\sin\theta_1 & \cos\theta_1 & 0 & 0 \\ 0 & 0 & \cos\theta_2 & \sin\theta_2 \\ 0 & 0 & -\sin\theta_2 & \cos\theta_2 \end{pmatrix} \right\}$$

is a maximal torus in $SO(4)$.

Proof: This clearly is a 2-torus and is a subgroup of $SO(4)$. As before, it suffices to prove that if $\phi \in SO(4)$ commutes with all elements of T then $\phi \in T$.

Let V be the 2-plane in R^4 spanned by e_1, e_2 and W be the 2-plane in R^4 spanned by e_3, e_4. We see that T consists of all pairs

$$(\text{rotation of } V, \text{ rotation of } W).$$

Claim: $\phi(e_1) \in V$.

Take $\alpha \in T$ such that α is the identity on V but is not the identity on W. Then

$$\phi(e_1) = ae_1 + be_2 + ce_3 + de_4$$

$$\alpha\phi(e_1) = ae_1 + be_2 + c'e_3 + d'e_4$$

$$\phi\alpha(e_1) = \phi(e_1) = ae_1 + be_2 + ce_3 + de_4 .$$

This shows $c = 0 = d$, so $\phi(e_1) \in V$. The same kind of proof shows $\phi(e_2) \in V$. Dually, $\phi(e_3)$ and $\phi(e_4)$ are in W.

So we know that ϕ is orthogonal on V and is orthogonal on W. A priori, it could be a reflection in each and we would still have $\phi \in SO(4)$. But ϕ commutes with all rotations on V and is thus a rotation on V. (See the proof of Proposition 3, Chapter V.) Similarly, ϕ is a rotation on W, so $\phi \in T$.

q.e.d.

From Propositions 1 and 2 the general result about maximal tori in $SO(n)$ should be clear. We have $n/2$ of the 2×2 rotation matrices for n even and have a 1 in the n,n position for n odd. The proof of the general case is an obvious extension of the above proofs.

Proposition 5:

$$T = \left\{ \begin{pmatrix} e^{i\theta_1} & & \bigcirc \\ & e^{i\theta_2} & \\ \bigcirc & & e^{i\theta_n} \end{pmatrix} \right\}$$

is a maximal torus in $U(n)$.

Proof: Let $\phi \in U(n)$ commute with each $\alpha \in T$. Consider any α of the form

$$\alpha = \begin{pmatrix} 1 & & \bigcirc \\ & e^{i\theta_2} & \\ \bigcirc & & e^{i\theta_n} \end{pmatrix} \in T .$$

Then $\alpha\phi(e_1) = \phi(\alpha(e_1)) = \phi(e_1)$. So α leaves $\phi(e_1)$ fixed, but such α's can move any vector which is not a multiple of e_1. Hence

$$\phi(e_1) = \lambda_1 e_1 \text{ , } \lambda_1 \in \mathbb{C} .$$

Similar arguments give

$$\phi(e_j) = \lambda_j e_j \quad \text{for} \quad j = 1,\ldots,n \ .$$

Thus

$$\phi = \begin{pmatrix} \lambda_1 & & & \bigcirc \\ & \lambda_2 & & \\ & & \ddots & \\ \bigcirc & & & \lambda_n \end{pmatrix}$$

and, since $\phi \in U(n)$, each λ_j is of unit length. Thus $\phi \in T$ and T is maximal.

Proposition 6:

$$T = \left\{ \begin{pmatrix} e^{i\theta_1} & & \bigcirc \\ & \ddots & \\ \bigcirc & & e^{i\theta_n} \end{pmatrix} \middle| \quad \theta_1 + \ldots + \theta_n = 0 \right\}$$

is a maximal torus in $SU(n)$.

Proof: A matrix α of the form $\begin{pmatrix} e^{i\theta_1} & & \bigcirc \\ & \ddots & \\ \bigcirc & & e^{i\theta_n} \end{pmatrix}$ has $\det \alpha = e^{i\theta_1} \ldots e^{i\theta_n} = e^{i(\theta_1+\ldots+\theta_n)}$ so that $\alpha \in SU(n) \Leftrightarrow \det \alpha = 1 \Leftrightarrow \theta_1 + \ldots + \theta_n = 0$. So the T described here is just the intersection of $SU(n)$ with the maximal torus given for $U(n)$.

First we must check that this is an $(n-1)$-torus. To do this, use

$$\begin{pmatrix} e^{i\theta_1} & & & \bigcirc \\ & \cdot & & \\ & & \cdot & \\ \bigcirc & & \cdot & \\ & & & e^{i\theta_n} \end{pmatrix} \leftrightarrow \begin{pmatrix} e^{i(\theta_1-\theta_n)} & & & \bigcirc \\ & e^{i(\theta_2-\theta_n)} & & \\ & & \ddots & \\ \bigcirc & & & e^{i(\theta_{n-1}-\theta_n)} \\ & & & & 1 \end{pmatrix}$$

$$\Sigma\theta_i = 0$$

It is an exercise to show that this works.

Now for $n > 2$ the same proof as used for $U(n)$ will work, but for $n = 2$

$$T = \begin{pmatrix} e^{i\theta} & 0 \\ 0 & e^{-i\theta} \end{pmatrix}$$

and we do not have matrices $\alpha = \begin{pmatrix} 1 & 0 \\ 0 & e^{-i\theta} \end{pmatrix}$ to use. But for $n = 2$ we give a direct simple proof. If

$$\phi = \begin{pmatrix} a & b \\ c & d \end{pmatrix} \in SU(2) \quad \text{and} \quad \alpha = \begin{pmatrix} i & 0 \\ 0 & -i \end{pmatrix} \in T$$

then

$$\phi\alpha = \begin{pmatrix} ai & -bi \\ ci & -di \end{pmatrix} = \begin{pmatrix} ai & bi \\ -ci & -di \end{pmatrix} = \alpha\phi \ .$$

Thus $b = 0 = c$ and $\phi \in T$.

<u>Proposition</u> 7: <u>The maximal torus given for</u> $U(n)$ <u>is also a maximal torus for</u> $Sp(n)$.

<u>Proof</u>: Just as for $U(n)$ we can show that any element of $Sp(n)$

which commutes with all $\begin{pmatrix} e^{i\theta_1} & & 0 \\ & \ddots & \\ 0 & & e^{i\theta_n} \end{pmatrix}$ must be diagonal with the diagonal elements having length 1 . But now these elements are quaternions. However (Exercise), any quaternion which commutes with i must be a complex number.

q.e.d.

C. Centers again

Now that we know maximal tori in our matrix groups, we are able to calculate the centers.

Proposition 8: Center $(Sp(n)) = \{I,-I\}$.

Proof: We have seen that if any element commutes with all elements of the maximal torus we have described, then it must lie in that maximal torus. Hence, in every case

$$\text{Center} \subset T .$$

If $A \in$ Center $Sp(n)$, then $A = \begin{pmatrix} e^{i\theta_1} & & 0 \\ & \ddots & \\ 0 & & e^{i\theta_n} \end{pmatrix}$. Since A must commute with the matrix jI , it follows that the diagonal elements must be real (and since they are of unit length) they are ± 1 . It is an exercise to show that they all have the same sign. Now I and $-I$ are in the center.

q.e.d.

<u>Proposition</u> 9: Center $U(n) = \{e^{i\theta}I\} \cong S^1$

Center $SU(n) = \{wI \mid w^n = 1\}$.

<u>Proof</u>: If $B \in$ Center $U(n)$ we get that B is diagonal with diagonal elements complex numbers of unit length. Let

$$B = \begin{pmatrix} \alpha_1 & & \bigcirc \\ & \alpha_2 & \\ & & \ddots \\ \bigcirc & & & \alpha_n \end{pmatrix} \quad \text{with each } |\alpha_\ell| = 1 \text{ . Let}$$

$$A = \begin{pmatrix} 0 & 1 & 0\ldots0 \\ 1 & 0 & 0\ldots0 \\ 0 & 0 & 1 \\ & \vdots & \quad .0 \\ & \vdots & 0 \ \ddots 1 \end{pmatrix} .$$

Then $AB = BA$ shows $\alpha_1 = \alpha_2$, etc., so all α are equal. Clearly any $e^{i\theta}I$ is in the center, so the center of $U(n)$ is as asserted.

For $SU(n)$ we note that the same argument will show that an element must be of the form $e^{i\theta}I$ to be in the center. But

$$\det(e^{i\theta}I) = e^{in\theta} ,$$

and since this must be 1 , $e^{i\theta}$ must be an nth root of unity.

q.e.d.

So Center $SU(n) = SU(n) \cap$ Center $U(n)$.

Finally, we want to calculate the center of $SO(n)$. It turns out that it depends on whether n is even or odd. The groups $SO(2n)$ and the groups $SO(2n+1)$ are different in some important ways.

Now $SO(2) = S^1$ is abelian so its center is the group itself. We have already proved that the center of $SO(3)$ is just $\{I\}$.

<u>Claim</u>: For $k \geq 3$ any element in the center of $SO(k)$ must be a diagonal matrix.

As before, if $A \in SO(k)$ is in the center, it must be in our standard maximal torus. So suppose $A \in$ Center $SO(k)$ is of the form

$$A = \begin{pmatrix} \cos\theta_1 & \sin\theta_1 & 0 \dots \\ -\sin\theta_1 & \cos\theta_1 & 0 \dots \\ 0 & 0 & * \dots \end{pmatrix}$$

Let

$$P = \begin{pmatrix} 0 & 0 & 1 & & \\ 0 & 1 & 0 & \bigcirc & \\ -1 & 0 & 0 & & \\ & & 1 & & 0 \\ & \bigcirc & & \ddots & \\ & & 0 & & 1 \end{pmatrix} \in SO(k) \quad .$$

Then PA has zero in the $1,2$ position, whereas AP has $\sin\theta_1$ in the $1,2$ position. Thus $\sin\theta_1 = 0$. Similar arguments show all off diagonal terms are zero.

It follows also (since each $\sin\theta_i = 0$) that each diagonal term ($\cos\theta_i$) is 1 or -1. So each 2×2 block is $\begin{pmatrix} 1 & 0 \\ 0 & 1 \end{pmatrix}$ or $\begin{pmatrix} -1 & 0 \\ 0 & -1 \end{pmatrix}$. Arguments like we used for $U(n)$ show that all diagonal terms must be equal. So we finally conclude that

$$\text{Center } SO(2n+1) = \{I\}$$

$$\text{Center } SO(2n) = \{I, -I\} \; .$$

For example, $\begin{pmatrix} -1 & & & \bigcirc \\ & -1 & & \\ & & -1 & \\ \bigcirc & & & -1 \end{pmatrix}$ is in the center of $SO(4)$, but $\begin{pmatrix} -1 & & \bigcirc \\ & -1 & \\ \bigcirc & & -1 \end{pmatrix}$ is not even in $SO(3)$.

We now tabulate the information we have generated about our groups.

Dimensions, Centers, Maximal tori

Group	Dimension	Center	Standard Maximal torus
$U(n)$	n^2	$\{e^{i\theta}I\} \cong S^1$	$\left\{\begin{pmatrix} e^{i\theta_1} & & \bigcirc \\ & \ddots & \\ \bigcirc & & e^{i\theta_n} \end{pmatrix}\right\}$
$SU(n)$	n^2-1	$\{wI \mid w^n=1\} \cong \mathbb{Z}_n$	$\left\{\begin{pmatrix} e^{i\theta_1} & & \bigcirc \\ & \ddots & \\ \bigcirc & & e^{i\theta_n} \end{pmatrix} \middle\vert\ \Sigma\theta_i = 0\right\}$
$SO(2n+1)$	$\frac{(2n+1)(2n)}{2} = 2n^2+n$	$\{I\}$	$\left\{\begin{pmatrix} \text{rot }\theta_1 & & \bigcirc \\ & \ddots & \\ \bigcirc & & \text{rot }\theta_n \; 1 \end{pmatrix}\right\}$
$SO(2n)$	$\frac{2n(2n-1)}{2} = 2n^2-n$	$\{I,-I\}$	$\left\{\begin{pmatrix} \text{rot }\theta_1 & & \bigcirc \\ & \ddots & \\ \bigcirc & & \text{rot }\theta_n \end{pmatrix}\right\}$ (for $n \geq 2$)
$Sp(n)$	$2n^2+n$	$\{I,-I\}$	$\left\{\begin{pmatrix} e^{i\theta_1} & & \bigcirc \\ & \ddots & \\ \bigcirc & & e^{i\theta_n} \end{pmatrix}\right\}$

Note that we have nothing to distinguish $SO(2n+1)$ and $\frac{Sp(n)}{\text{Center}}$. A good part of the remainder of this book is devoted to deciding for which n these are isomorphic.

D. Exercises

1. Show that the operation defined on $G \times H$ does make it into a group. Prove that if G and H are abelian, so is $G \times H$.

2. Do the exercise in the first example of §A.

3. Show directly that the product of the matrices

$$\begin{pmatrix} \cos\theta & \sin\theta \\ -\sin\theta & \cos\theta \end{pmatrix} \text{ and } \begin{pmatrix} \cos\phi & \sin\phi \\ -\sin\phi & \cos\phi \end{pmatrix}$$

is the matrix for a rotation through angle $\theta + \phi$.

4. Let T be a maximal torus in a matrix group G and let $x \in G$. Prove that xTx^{-1} is also a maximal torus in G.

5. Prove that if q is a quaternion such that $qi = iq$, then q is a complex number.

6. Show that $\dfrac{U(n)}{\text{center}} \cong \dfrac{SU(n)}{\text{center}}$.

7. A <u>lattice subgroup</u> K of $\mathbb{R}^n$ consists of all integar linear combinations of some set of linearly independent vectors. More explicitely, let

$$v_1, v_2, \ldots, v_k$$

be linearly independent vectors in $\mathbb{R}^n$. For any integers $a_1, \ldots, a_k$ we form

$$a_1v_1 + \ldots + a_kv_k .$$

Then $K = \{a_1v_1 + \ldots + a_kv_k\} \mid a_i \in \mathbb{Z}\}$. It is routine to verify that K is a subgroup of $\mathbb{R}^n$.

A subgroup H of $\mathbb{R}^n$ is <u>discrete</u> if some neighborhood of 0 in $\mathbb{R}^n$ contains no point of H other than 0. Prove that: A discrete subgroup of $\mathbb{R}^n$ is a lattice subgroup. (First let H be a discrete subgroup of $\mathbb{R}$. Choose $h_o \in H$ with $h_o \neq 0$ but with no elements of H between 0 and h_o. We call the closed interval $[0,h_o]$ a <u>fundamental domain</u> D for H. Note that $\{0,h_o\}$ are the only elements of H in D. We claim that $H = \mathbb{Z}h_o$. Clearly $\mathbb{Z}h_o \subset H$. So suppose $\exists\ h \notin \mathbb{Z}h_o$ in H. Then h lies between some nh_o and $(n+1)h_o$

$$\underset{0}{+}\overset{D}{\rule{4em}{0.4pt}}\underset{h_o}{+} \qquad \underset{nh_o}{+}\overset{h}{\bullet}\underset{(n+1)h_o}{+} .$$

But then $h - nh_o$ is an element of H in D different from 0 and different from h_o. This contradicts our choice of h_o.

Next, for $\mathbb{R}^2$ (with H a discrete subgroup) we choose $\alpha,\beta \in H$ with α,β linearly independent and with the property that the closed parallelogram D which α,β span contains no elements of H other than $0,\alpha,\beta,\alpha+\beta$.

The proof that

$$H = \{a\alpha + b\beta \mid a,b \in \mathbb{Z}\}$$

is similar to our proof for $\mathbb{R}$.)

8. Show that if L is a lattice group in $\mathbb{R}^n$ generated by $v_1,\ldots,v_k$. Then $\mathbb{R}^n/L$ is isomorphic with the product of k-torus and $\mathbb{R}^{n-k}$.

Chapter 8
Covering by Maximal Tori

A. General remarks

In Exercise 4 of Chapter VII one showed that if T is a maximal torus in a matrix group G, then for any $x \in G$, xTx^{-1} is also a maximal torus. What we prove in this chapter is that if T is our standard maximal torus in one of our connected matrix groups G, then

$$(\dagger) \qquad G = \bigcup_{x \in G} xTx^{-1}$$

showing that every element of G lies in at least one maximal torus.

To say that $G = \bigcup_{x \in G} xTx^{-1}$ is to say that given $y \in G$ there exists $x \in G$ such that

$$y \in xTx^{-1} .$$

This is equivalent to: Given $y \in G$, there exists $z \in G$ such that $zyz^{-1} \in T$ (take $x = z^{-1}$). So we want to show that $y \in G$ can be put in diagonal or 2×2 block-diagonal form by conjugation.

We begin by reviewing a little linear algebra. Let V be a vector space over a field k and let

$$\phi : V \to V$$

be a linear map.

Definition: A subspace W of V is ϕ - stable if $\phi(W) \subset W$. In this case, we can restrict the domain of ϕ to W to get a linear map

$$\phi|W : W \to W \quad .$$

Examples: Let $V = \mathbb{R}^3$ and W be the 2-plane spanned by e_1 and e_2. If

$$\phi = \begin{pmatrix} \cos\theta & \sin\theta & 0 \\ -\sin\theta & \cos\theta & 0 \\ 0 & 0 & 1 \end{pmatrix}$$

then W is ϕ - stable.

Let $V = \mathbb{R}^2$ and $\phi = \begin{pmatrix} 0 & 1 \\ 1 & 0 \end{pmatrix}$. Then $W = \{(x,x)\}$ is ϕ - stable and so is $W' = \{(x,-x)\}$.

An important special case occurs when $\dim W = 1$.

Definition: A nonzero vector $v \in V$ is an eigenvector for ϕ if there exists a $\lambda \in k$ such that

$$\phi(v) = \lambda v \quad .$$

λ is the corresponding eigenvalue.

Now if ϕ has one eigenvector v then it has a "line" of them; namely, if $r \in k$ with $r \neq 0$, then

$$\phi(rv) = r\phi(v) = r\lambda v = \lambda(rv) \quad .$$

(We will have to be a little more careful when using our "skew" field $\mathbb{H}$.) Thus rv is also an eigenvector with the same eigenvalue λ. This suggests that eigenvalues may be more fundamental.

Definition: For any $\lambda \in k$, set

$$V(\lambda) = \{v \in V \mid \phi(v) = \lambda v\} \quad .$$

Thus $V(\lambda)$ includes the zero vector o and all eigenvectors having eigenvalue λ . It is easy to see that $V(\lambda)$ is a subspace of V , the **eigenspace** belonging to λ .

For example, taking $\lambda = 0$, we see that

$$V(0) = \{v \in V \mid \phi(v) = 0\}$$

is just the kernel (or null space) of the linear map ϕ .

Now take $\dim V$ to be finite. We may as well take $V = k^n$. Then linear maps $\phi : V \to V$ correspond 1-1 with elements of $M_n(k)$, so we can think of $\phi \in M_n(k)$.

Proposition 1: $V(\lambda) \neq \{o\} \Leftrightarrow \det(\phi - \lambda I) = 0$.

Proof: $\Rightarrow$

If $o \neq v$ satisfies $\phi(v) = \lambda v$, then $(\phi - \lambda I)v = o$ showing that $\phi - \lambda I$ is singular. Thus $\det(\phi - \lambda I) = 0$.

$\Leftarrow$

If $\det(\phi - \lambda I) = 0$, then $\phi - \lambda I$ must send some nonzero vector v to o ; i.e., $\phi(v) = \lambda v$, so $V(\lambda) \neq \{o\}$.

B. (†) **for** U(n) **and** SU(n)

We are concerned with C - linear maps $C^n \to C^n$ which we represent by elements of $M_n(\mathbb{C})$. We begin with two easy observations:

(a) If A is unitary and $W \subset \mathbb{C}^n$ is A - stable, then $A|W$ is unitary.

If $x,y \in W$ then xA and $yA \in W$ and

$$\langle xA, yA \rangle = \langle x,y \rangle .$$

(b) If A is unitary and $W \subset \mathbb{C}^n$ is A - stable, then the orthogonal complement

$$W^{\perp} = \{x \in \mathbb{C}^n \mid \langle x,y \rangle = 0 \quad \text{for all} \quad y \in W\}$$

is also A - stable.

Let $x \in W^{\perp}$. For any $y \in W$ we have

$$\langle xA, y \rangle = \langle x, y{}^t\bar{A} \rangle = \langle x, yA^{-1} \rangle \quad .$$

Now A is an isomorphism of W onto W so $yA^{-1} \in W$. Thus

$$\langle xA, y \rangle = 0 \quad \text{for all} \quad y \in W , \quad \text{showing} \quad xA \in W^{\perp} \quad .$$

<u>Proposition</u> 2: <u>For</u> $A \in U(n)$ <u>there exists an orthonormal basis of eigenvectors</u> $v_1 ,\ldots, v_n \in \mathbb{C}^n$ <u>for</u> A .

<u>Proof</u>: By (a) and (b) above, it suffices to show that a unitary matrix C always has an eigenvector v . For we can make v unit length and restrict C to $v^{\perp}$, etc.

But let $A \in M_n(\mathbb{C})$. Then

$$p(\lambda) = \det(A - \lambda I)$$

is a polynomial in λ and $\mathbb{C}$ is algebraically closed so it has a root. Thus A has an eigenvector.

<u>Proposition</u> 3: <u>The conjugates of</u> $T = \begin{pmatrix} e^{i\theta_1} & & \bigcirc \\ & \ddots & \\ \bigcirc & & e^{i\theta_n} \end{pmatrix}$ <u>cover</u> $U(n)$.

<u>Proof</u>: Given $A \in U(n)$ we will find $B \in U(n)$ such that

$$BAB^{-1} \in T .$$

Let $v_1, \ldots, v_n$ be an orthonormal basis consisting of eigenvectors of A . Let B send

$$e_j \text{ to } v_j .$$

Then BAB^{-1} sends e_j to v_j to $\lambda_j v_j$ to $\lambda_j e_j$. So

$$BAB^{-1} = \begin{pmatrix} \lambda_1 & & & \bigcirc \\ & \lambda_2 & & \\ & & \ddots & \\ \bigcirc & & & \lambda_n \end{pmatrix} .$$

Since BAB^{-1} is unitary, each $\lambda_j e_j$ is a unit vector; i.e., $|\lambda_j| = 1$. Thus $BAB^{-1} \in T$. q.e.d.

<u>Corollary</u>: <u>The conjugates of</u>

$$T = \left\{ \begin{pmatrix} e^{i\theta_1} & & \bigcirc \\ & \ddots & \\ \bigcirc & & e^{i\theta_n} \end{pmatrix} \middle| \ \Sigma\theta_j = 0 \right\}$$

<u>cover</u> $SU(n)$.

<u>Proof</u>: Given $A \in SU(n)$ we want $B \in SU(n)$ such that $BAB^{-1} \in T$. If we take $B' \in U(n)$ such that

$$B'A(B')^{-1} \text{ is in the maximal torus for } U(n) ,$$

it is actually in the maximal torus for $SU(n)$ since

$$\det(B'A(B')^{-1}) = \det A = 1 .$$

If we choose $\mu \in C$ such that $\det B' = \frac{1}{\mu^n}$ and set

$$B = \mu B' ,$$

then $BAB^{-1} = B'A(B')^{-1}$ and $B \in SU(n)$.

C. (†) <u>for</u> SO(n)

For any matrix group G and maximal torus T we have that

$$\bigcup_{x \in G} xTx^{-1}$$

is a connected set, so if G is not connected, (†) could not hold. Since $\mathcal{O}(n)$ is not connected, we know that

$$\mathcal{O}(n) \neq \bigcup_{x \in \mathcal{O}(n)} xTx^{-1} .$$

But we will prove (†) for $SO(n)$ and this will, incidentally, prove that

$$SO(n) \text{ is connected} .$$

Clearly we can have no result like Proposition 2 for $SO(n)$. For example $A = \begin{pmatrix} \cos\theta & \sin\theta \\ -\sin\theta & \cos\theta \end{pmatrix} \in SO(2)$ with $A \neq I$ has no nonzero eigenvector. But this is about all that can happen--we can find stable 2-planes. To see this we consider symmetric linear maps.

Definition: A linear map $S : \mathbb{R}^n \to \mathbb{R}^n$ is symmetric if

$$\langle xS, y\rangle = \langle x, yS\rangle \quad \text{for all} \quad x, y \in \mathbb{R}^n .$$

Note that for $A \in M_n(\mathbb{R})$, $S = A + {}^tA$ is symmetric. For we have

$$\langle xS, y\rangle = \langle xA, y\rangle + \langle x\,{}^tA, y\rangle = \langle x, y\,{}^tA\rangle + \langle x, yA\rangle = \langle x, Sy\rangle .$$

Corresponding to (a) and (b) for $U(n)$ we have: If S is symmetric and $W \in \mathbb{R}^n$ is S-stable, then $S|W$ is symmetric and $W^{\perp}$ is S-stable. Thus, just as for $U(n)$, it suffices to show that a symmetric matrix always has an eigenvector.

Proposition 4: A (real) symmetric matrix S has an eigenvector.

Proof: Define $\Phi : \mathbb{R}^n - (0) \to \mathbb{R}$ by

$$\Phi(x) = \frac{\langle x, xS\rangle}{\langle x, x\rangle} .$$

Note that for nonzero $r \in \mathbb{R}$ we have $\Phi(rx) = \Phi(x)$. Thus if S^{n-1} is the unit sphere, we have

$$\Phi(\mathbb{R}^n - (0)) = \Phi(S^{n-1}) .$$

Now Φ is continuous and S^{n-1} is compact. Thus $\Phi(S^{n-1})$ is compact and we can find $v \in S^{n-1}$ such that

$\Phi(v)$ is an <u>absolute</u> <u>maximum</u> .

We claim that v is then an eigenvector for S .

For any nonzero $y \in R^n$, if we set

$$f(t) = \Phi(v + ty) ,$$

we must have $f'(0) = 0$. This will allow us to show v is an eigenvector and to find its eigenvalue.

We have

$$f(t) = \frac{\langle v+ty, vS+tyS\rangle}{\langle v+ty, v+ty\rangle} = \frac{\alpha(t)}{\beta(t)} ,$$

$$f'(0) = \frac{\alpha'(0)\beta(0) - \beta'(0)\alpha(0)}{(\beta(0))^2} .$$

Since v is of unit length, $\beta(0) = 1$ so that

$$f'(0) = \alpha'(0) - \beta'(0)\alpha(0) .$$

We easily calculate that $\beta'(0) = 2\langle v,y\rangle$ and $\alpha(0) = \langle v,vS\rangle$ and $\alpha'(0) = \langle y,vS\rangle + \langle v,yS\rangle = 2\langle vS,y\rangle$ (S and $\langle\ ,\ \rangle$ are symmetric). This gives $f'(0) = 2\langle vS,y\rangle - 2\langle v,vS\rangle\langle v,y\rangle = 0$, or

$$\langle(vS - \langle v,vS\rangle v),y\rangle = 0 .$$

Now $\langle\ .\ \rangle$ is nondegenerate and y is arbitrary, so

$$vS = \langle v.vS\rangle v .$$ q.e.d.

We first use Proposition 4 to study $\mathfrak{G}(n)$ and then prove (+) for $SO(n)$.

Proposition 5: Let $A \in \mathfrak{O}(n)$; then there exists an A-stable subspace W of $\mathbb{R}^n$ with

$$\dim W \in \{1,2\} \quad .$$

Proof: $S = A + {}^tA$ is symmetric. Let w be an eigenvector for S . Consider w and wA .

case (i) If w and wA are linearly dependent, $wA = \lambda w$, then $W = \{rw \mid r \in \mathbb{R}\}$ is A-stable and $\dim W = 1$.

case (ii) If linearly independent, let $W = \mathrm{Span}(w,wA)$. We have $wS = \lambda w$, i.e., $wA + w{}^tA = \lambda w$ and ${}^tA = A^{-1}$ so that

$$wA + wA^{-1} = \lambda w \quad \text{or} \quad wA^2 + w = \lambda wA \ .$$

This implies that W is A-stable. For, let

$$\alpha w + \beta wA \in W \ .$$

Then

$$\begin{aligned}(\alpha w + \beta wA)A &= \alpha wA + \beta wA^2 \\ &= \alpha wA + \beta(\lambda wA - w) \\ &= (-\beta)w + (\alpha+\beta\lambda)wA \in W \quad .\end{aligned}$$

Proposition 6: The conjugates of our standard maximal torus in $SO(n)$ cover $SO(n)$.

Proof: We choose an orthonormal basis $v_1,\ldots,v_n$ using $A \in SO(n)$ as follows.

If A has a stable one-dimensional subspace, choose one and take

v_1 to be a unit vector in it. Next do the same for $v_1^{\perp}$, and continue as long as possible, generating an orthonormal set $v_1, \ldots, v_k$ of eigenvectors for A. Let $W = \mathrm{span}(v_1, \ldots, v_k)^{\perp}$. A is orthogonal on W. Choose a stable 2-space W_1 and let v_{k+1}, v_{k+2} be an orthonormal basis for W_1, etc. This gives an orthonormal basis

$$v_1, v_2, \ldots, v_k, \underbrace{v_{k+1}, v_{k+2}}_{W_1}, \ldots, \underbrace{v_{n-1}, v_n}_{W_{\frac{n-k}{2}}} .$$

We use this to find $B \in SO(n)$ such that

$$BAB^{-1} \in T = \text{our standard maximal torus.}$$

Let C map e_i to v_i. Then $C \in \mathcal{O}(n)$ and

$$CAC^{-1} \in T .$$

(Note that $\det CAC^{-1} = \det A = 1$.) If it happens that $\det C = 1$, let $B = C$. If $\det C = -1$ we seek $D \in \mathcal{O}(n) - SO(n)$ such that $DTD^{-1} = T$. Then $B = DC$ will satisfy $BAB^{-1} \in T$ and $\det B = 1$. Such a D is easy to find; e.g.,

$$D = \begin{pmatrix} 0 & 1 & & & \\ 1 & 0 & & \bigcirc & \\ & & 1 & & \\ & \bigcirc & & \ddots & \\ & & & & 1 \end{pmatrix} .$$

This proves $(+)$ for $SO(n)$.

D. (t) <u>for</u> Sp(n)

<u>Proposition</u> 7: Given $A \in Sp(n), \exists\, v \in \mathbb{H}^n$ such that the span of v and jv is A-stable.

Proof: Using the map ψ defined on page 18, we can make Sp(n) into a subgroup of SU(2n). So we can consider $A \in SU(2n)$ (and identify $\mathbb{H}^n = \mathbb{C}^{2n}$). Then we can find $v \in \mathbb{H}^n$ such that

$$vA = \lambda v \quad \text{with} \quad \lambda \in \mathbb{C}.$$

Then $(jv)A = j(vA) = j\lambda v = \bar{\lambda} jv$. Thus the $\mathbb{H}$-span of v (= span (v, jv)) is A-stable.

<u>Proposition</u> 8: (+) <u>holds</u> <u>for</u> Sp(n) .

<u>Proof</u>: Just as for U(n) we now find an orthonormal basis of eigenvectors and thus $B \in Sp(n)$ such that

$$BAB^{-1} = \begin{pmatrix} \lambda_1 & & 0 \\ & \ddots & \\ 0 & & \lambda_n \end{pmatrix}$$

with the $\lambda_1, \ldots, \lambda_n$ quarternions of unit length. If we conjugate this by a diagonal matrix

$$Q = \begin{pmatrix} q_1 & & \bigcirc \\ & \ddots & \\ \bigcirc & & q_n \end{pmatrix} \in Sp(n) \ , \quad |q_j| = 1 \ ,$$

we get

$$QBA(QB)^{-1} \quad .$$

This will be in the standard maximal torus if each $q_j \lambda_j q_j^{-1} = q_j \lambda_j \bar{q}_j$ is a complex number. It is proved in Exercise 5 that we can choose the q_j's to do this.

E. Relfections in R^n (again)

Let $x \in R^n$ be nonzero and let $x^{\perp}$ be the (n-1)-hyperplane through o perpendicular to x .

Proposition 9: The reflection $\phi : R^n \to R^n$ in $x^{\perp}$ is given by the formula

$$\phi(y) = y - 2 \frac{\langle x,y \rangle}{\langle x,x \rangle} x \quad .$$

Proof:

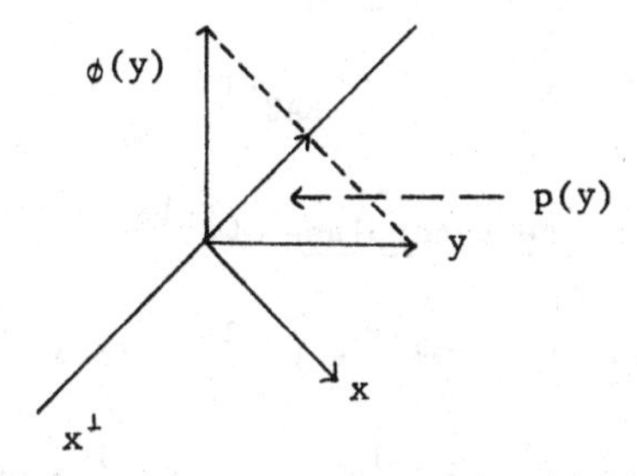

Let $p(y)$ be the projection of y in $x^\perp$. Then $y - p(y)$ is some multiple of x, say rx, and $\langle p(y), x\rangle = 0$. This gives $r = \frac{\langle x,y\rangle}{\langle x,x\rangle}$. So

$$p(y) = y - \frac{\langle x,y\rangle}{\langle x,x\rangle} x \quad \text{and} \quad \phi(y) - 2\frac{\langle x,y\rangle}{\langle x,x\rangle} x \ .$$

We can check directly that ϕ is an orthogonal map.

$$\langle \phi(y), \phi(z)\rangle = \langle y - 2\frac{\langle x,y\rangle}{\langle x,x\rangle} x,\ z - 2\frac{\langle x,z\rangle}{\langle x,x\rangle} x\rangle$$

$$= \langle y,z\rangle - 2\frac{\langle x,z\rangle}{\langle x,x\rangle}\langle y,x\rangle - 2\frac{\langle x,y\rangle}{\langle x,x\rangle}\langle x,z\rangle + 4\frac{\langle x,y\rangle\langle x,z\rangle}{\langle x,x\rangle\langle x,x\rangle}\langle x,x\rangle$$

and this is just $\langle y,z\rangle$.

Also it is easy to see that $\det \phi = -1$ because we can choose a basis with x as first vector and the other vectors forming a basis for $x^\perp$. Relative to this basis ϕ is given by

$$\begin{pmatrix} -1 & & \bigcirc \\ & 1 & \\ \bigcirc & & 1 \end{pmatrix} .$$

<u>Proposition</u> 10: *If $A \in \mathcal{O}(n)$ is reflection in a hyperplane W in R^n and $B \in \mathcal{O}(n)$, then*

$$BAB^{-1}$$

is reflection in the hyperplane WB^{-1}.

<u>Proof</u>: Let $w' = wB^{-1} \in WB^{-1}$. Then

$$w'BAB^{-1} = wB^{-1}BAB^{-1} = wAB^{-1} = wB^{-1} = w' ,$$

so that BAB^{-1} is the identity on WB^{-1}.

Let $\alpha_1,\ldots,\alpha_n$ be an orthonormal basis with α_1 perpendicular to W and $\alpha_2,\ldots,\alpha_n$ a basis fpr W. Then BAB^{-1} is the identity on $\alpha_2B^{-1},\ldots,\alpha_nB^{-1}$ and

$$(\alpha_1B)(B^{-1}AB) = \alpha_1AB = -\alpha_1B ,$$

proving the proposition.

Proposition 11: $\mathfrak{G}(n)$ is generated by reflections.

Proof: We want to show that any element of $\mathfrak{G}(n)$ may be written as the product of a finite number of reflections.

First we prove this for elements of our standard torus T. Let m be the biggest integer $\leq \frac{n}{2}$ so that T has m blocks $B_1 = \begin{pmatrix} \cos\theta_1 & \sin\theta_1 \\ -\sin\theta_1 & \cos\theta_1 \end{pmatrix}, \ldots, B_m = \begin{pmatrix} \cos\theta_m & \sin\theta_m \\ -\sin\theta_m & \sin\theta_m \end{pmatrix}$ arranged along the diagonal. Let $\psi_i \in T$ be the identity except in block B_i. Then any element of T is surely a finite product of elements $\psi_1,\ldots,\psi_m$.

Let ϕ_i be reflection in the hyperplane formed by all coordinates except those for B_i and the first coordinate for B_i.

Let ρ_i be reflection in

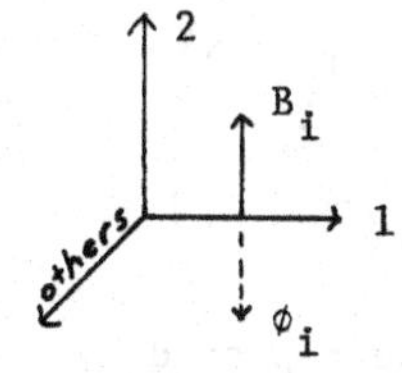

the hyperplane formed by all coordinates except those for B_i and the coordinate at angle $\frac{\theta_i}{2}$ in the B_i plane. Then $\rho_i \circ \phi_i = \psi_i$. Thus every element of T is the product of finitely many reflections.

Given $A \in SO(n)$, choose B such that

$$BAB^{-1} \in T \ .$$

Writing BAB^{-1} as a finite product of reflections and using Proposition 12 gives A as such a product. Finally, let $C \in \mathcal{O}(n)$. Then $C = AD$ where $A \in SO(n)$ and D is a reflection (Exercise).

q.e.d.

F. Exercises

1. Show that $A = \begin{pmatrix} \cos\frac{\pi}{4} & \sin\frac{\pi}{4} \\ -\sin\frac{\pi}{4} & \cos\frac{\pi}{4} \end{pmatrix} \in SO(2)$ has no A - stable subspace in $\mathbb{R}^2$ other than $\{o\}$. Find all A - stable subspaces in $\mathbb{R}^2$ for $A = \begin{pmatrix} 0 & 1 \\ 1 & 0 \end{pmatrix}$.

2. Find all eigenvectors and eigenvalues of

$$\begin{pmatrix} 0 & 1 \\ 1 & 0 \end{pmatrix}, \begin{pmatrix} 1 & 0 \\ 0 & 0 \end{pmatrix}, \begin{pmatrix} 1 & 0 \\ 0 & 1 \end{pmatrix}, \begin{pmatrix} 0 & 1 \\ 0 & 1 \end{pmatrix} .$$

3. In $\mathbb{H}^n$ show that if $\langle x,y \rangle$ has zero real part for every y, then $x = 0$.

4. Let $D = \begin{pmatrix} -1 & & & \bigcirc \\ & 1 & & \\ & & \ddots & \\ \bigcirc & & & 1 \end{pmatrix}$. Show that $\mathcal{O}(n) - SO(n) = SO(n)D$.

Show that D may be replaced by any other element of $\mathcal{O}(n) - SO(n)$.

5. Prove that if $\lambda \in \mathbb{H}$ has length 1, there exists $q \in \mathbb{H}$ of length 1 such that $q\lambda\bar{q}$ is a complex number.

6. Let L_1, L_2 be two lines through o in $\mathbb{R}^2$ with angle θ between them. Let ϕ_1 be reflection in L_1 and ϕ_2 be reflection in L_2. Let W be the subgroup of $\mathfrak{O}(2)$ generated by ϕ_1 and ϕ_2. For what values of θ will W be a finite group?

7. Recall that in order to prove that

$$1 + \dim SU(n) = \dim U(n)$$

in Chapter IV we used the fact that

$$e^{TrA} = \det e^A \quad .$$

Prove this now for $A \in U(n)$. (We can write

$$A = BCB^{-1}$$

where B is unitary and C is an element of our standard maximal torus in U(n). Then

$$e^A = e^{BCB^{-1}} = Be^C B^{-1}$$

and $C = \begin{pmatrix} \lambda_1 & & \bigcirc \\ & \ddots & \\ \bigcirc & & \lambda_n \end{pmatrix}$ with $|\lambda_j| = 1$ so $e^C = \begin{pmatrix} e^{\lambda_1} & & \bigcirc \\ & \ddots & \\ \bigcirc & & e^{\lambda_n} \end{pmatrix}$.

Thus $\det e^A = \det (Be^C B^{-1}) = \det e^C = e^{\lambda_1} \dots e^{\lambda_n} = e^{\lambda_1 + \dots + \lambda_n}$ $= e^{TrC} = e^{TrA}$.)

Chapter 9
Conjugacy of Maximal Tori

A. Monogenic groups

Definition: A subset Y of a space X is said to be dense in X if every nonempty open set in X contains at least one point in Y.

Examples: Both $\mathbb{Q}$ and $\mathbb{R} - \mathbb{Q}$ are dense in $\mathbb{R}$. $\{(x_1,x_2) \in \mathbb{R}^2 \mid x_1,x_2 \in \mathbb{Q}\}$ is dense in $\mathbb{R}^2$.

Lemma: If Y is dense in X and C is a closed set in X with $Y \subset C$, then $C = X$.

Proof: $X - C$ is open, so if it is nonempty it must contain some $y \in Y$. But $y \in X - C$ contradicts $Y \subset C$.

Definition: A matrix group G is monogenic (i.e., one generator) if there exists $x \in G$ such that

$$\{x, x^2, x^3, \ldots\}$$

is dense in G . Then any such x is called a generator.

Note that the additive group $\mathbb{R}$ is not monogenic, nor are any of the vector groups $\mathbb{R}^n$. Of course, the identity (matrix) cannot ever be a generator (if $G \neq I$). Consider the circle group S^1 thought of as the additive group of reals modulo 1 . If $x = \frac{p}{q} \in S^1$ is rational, then the powers

$$x, 2x, 3x, 4x, \ldots$$

will not form a dense set in S^1 because they will all lie in the set

$$\{\frac{1}{q}, \frac{2}{q}, \ldots, \frac{q-1}{q}, \frac{q}{q} = 0\} \quad .$$

But if we choose x to be irrational we should get a generator. However, we will use a more topological way of finding a generator, and it will generalize immediately to higher dimensional tori.

Proposition 1: S^1 is monogenic (S^1 = reals mod 1) .

Proof: Let $U_1, U_2, U_3, \ldots$ be a countable basis for the open sets in S^1 . Then to prove the proposition it suffices to find $x \in S^1$ such that: Given U_k , there exists an $n \in N = \{1,2,3,\ldots\}$ such that

$$nx \in U_k \; .$$

We proceed as follows.

Choose a nondegenerate closed interval $I_1 = [a_1, b_1] \subset U_1$.

Choose an $n_1 \in N$ such that length $[n_1 a_1, n_1 b_1] \geq 2$. Then $n_1 I_1$ is all of S^1 , and thus we can choose I_2 (nondeg.) such that $n_1 I_2 \subset U_2$,

choose n_2 such that $n_2 I_2$ is all of S^1.

Then we can

choose I_3 (non deg.) such that $n_2 I_3 \subset U_3$,

choose n_3 such that $n_3 I_3$ is all of S^1 , etc.

Because $U_1, U_2, U_3, \ldots$ must contain arbitrarily small sets, we have that $I_1 \cap I_2 \cap I_3 \cap \ldots$ is a single element x .

But x is a generator for S^1 . For, given U_k , we have $n_{k-1} x \in n_{k-1} I_k \subset U_k$. q.e.d.

Now the r-dimensional torus T is just all r-tuples of real numbers, each taken modulo 1 . So by using cubes instead of intervals, we get

Proposition 2: *The r-torus T is monogenic.*

B. *Conjugacy of maximal tori*

We are now ready to prove that for a connected matrix group, any two maximal tori are conjugate. Let $G \in \{SO(n), U(n), SU(n), Sp(n)\}$ and let T be our standard maximal torus in G . We know that

$$G = \bigcup_{x \in G} xTx^{-1} .$$

Proposition 3: *Let T' be any maximal torus in G . Then for some $x \in G$, $T' = xTx^{-1}$.*

Proof: Since T' is a torus it is monogenic and we choose a generator y for T' . Then for some x

$$y \in xTx^{-1} .$$

Now xTx^{-1} is a group, so

$$Y = \{y, y^2, y^3, \ldots\} \subset xTx^{-1} ,$$

and, of course, $Y \subset T'$. Thus Y is contained in the closed set $T' \cap xTx^{-1}$ in T' and Y is dense in T' . So by our lemma we have

$$T' \cap xTx^{-1} = T' \quad \text{or} \quad xTx^{-1} \supset T' .$$

But xTx^{-1} is a maximal torus in G so $xTx^{-1} = T'$.

Definition: The rank of a matrix group G is the dimension of a maximal torus in G .

This is clearly an invariant for isomorphisms.

C. The isomorphism question again

For a matrix group G we now have two numerical invariants, its dimension n and its rank r , and one subgroup invariant, its center C . Furthermore, we have calculated these and we now tabulate our results according to rank.

Rank 1:	Group	Dimension	Center
	U(1)	1	S^1
	SU(2)	3	$\mathbb{Z}/2$
	SO(2)	1	S^1
	SO(3)	3	$\{I\}$
	Sp(1)	3	$\{I,-I\}=\mathbb{Z}/2$.
Rank 2:			
	U(2)	4	S^1
	SU(3)	8	$\mathbb{Z}/3$
	SO(4)	6	$\{I,-I\}=\mathbb{Z}/2$
	SO(5)	10	$\{I\}$
	Sp(2)	10	$\{I,-I\}$.
Rank 3:			
	U(3)	9	S^1
	SU(4)	15	$\mathbb{Z}/4$
	SO(6)	15	$\{I,-I\}=\mathbb{Z}/2$
	SO(7)	21	$\{I\}$
	Sp(3)	21	$\{I,-I\}=\mathbb{Z}/2$.

For rank 4 and greater we have

$$\dim U < \dim SU < \dim SO(\text{even}) < \dim SO(\text{odd}) = \dim Sp .$$

Since SO(odd) and Sp have different centers, they are not isomorphic. So we just need to look at ranks 1, 2, 3.

We know that $U(1) \cong SO(2)$ (circle group). $SO(3) \not\cong Sp(1)$ and we proved in Chapter II that $Sp(1) \cong SU(2)$. So we know all about rank 1.

For rank 2 only SO(5) and Sp(2) have the same dimension and they have nonisomorphic centers.

For rank 3 SU(4) and SO(6) have the same dimension and different centers, and the same applies to SO(5) and Sp(2) . So we have solved our isomorhism problem for these groups. But we can generate some other groups and the isomorphism question for these needs to be resolved. In the next section we consider ways of getting new groups.

If H is any proper subgroup of the center (S^1) of U(n) , then we have a group U(n)/H , and H is finite, so

$$\dim \frac{U(n)}{H} = n^2 = \dim U(n)$$

so these groups are not isomorphic to any of our others.

If H is a finite proper subgroup of the center $\mathbb{Z}/n$ of SU(n) (n not a prime), then

$$\dim \frac{SU(n)}{H} = n^2 - 1 = \dim SU(n)$$

and again these are not isomorphic with any of our other groups except we might have

$$\frac{SU(4)}{\mathbb{Z}/2} \simeq SO(6) .$$

Indeed, we will see later that this is the case.

D. Simple groups, simply-connected groups

Definition: A matrix group G is simple if it has no nontrivial normal subgroup. This is the same as saying it has no quotient groups other than itself and the trivial group.

If G is not simple, then by choosing a nontrivial normal subgroup H of G we get a new group G/H. So we want to see which of our groups are simple, and, when they are not, what normal subgroups they have. We state the result, but a proof would take us too far afield (See [3]).

Theorem: $\frac{U(n)}{\text{center}} \cong \frac{SU(n)}{\text{center}}$ *is simple*.

$SO(2n+1)$ *is simple*.

$\frac{SO(2n)}{\text{center}}$ *is simple*.

$\frac{Sp(n)}{\text{center}}$ *is simple*.

So we only get the groups we already know about.

Another way of generating new groups is somewhat more sophisticated. A *path* ω in G is a smooth curve $\omega : [0,1] \to G$ with $\omega(0) = e$. Let $P(G)$ be the set of all paths in G. Then $P(G)$ becomes a group if we define

$$(\omega\sigma)(t) = \omega(t)\sigma(t) .$$

Proposition 4: *If* G *is a connected group, the map* $\rho : P(G) \to G$, *defined by*

$$\rho(\omega) = \omega(1)$$

is a surjective homomorphism (See [3]).

Proof: G connected implies that for any $x \in G$ there is a path from e to x in G, so ρ is surjective. Now

$$\rho(\omega\sigma) = (\omega\sigma)(1) = \omega(1)\sigma(1) = \rho(\omega)\rho(\sigma) ,$$

so ρ is a homomorphism.

We denote the kernel of ρ by $\Omega(G)$ and call it the loops in G.

Definition: G is simply-connected means that $\Omega(G)$ is connected.

If G is not simply-connected, let $\Omega^\circ(G)$ be the identity component of $\Omega(G)$. Then $\Omega^\circ(G)$ is also a normal subgroup of $P(G)$ and we set

$$\tilde{G} = \frac{P(G)}{\Omega^\circ(G)}$$

and call this group the universal covering group of G.

So we get new groups this way whenever G is not simply-connected. Clearly $\rho : P(G) \to G$ induces a homomorphism

$$\rho : \tilde{G} \to G .$$

The kernel of ρ is denoted by $\pi_1(G)$ and is called the fundamental group of G. Clearly G is simply-connected $\Leftrightarrow \pi_1(G) = 1$ (i.e., the trivial group). We state but do not prove the following.

<u>Theorem</u>: $\pi_1(Sp(n)) = 1$ for $n = 1,2,\ldots$

$\pi_1(U(n)) = \mathbb{Z}$ for $n = 2,3,\ldots$

$\pi_1(SU(n)) = 1$ for $n = 2,3,\ldots$

$\pi_1(SO(n)) = \frac{\mathbb{Z}}{2}$ for $n = 3,4,\ldots$.

We have $\widetilde{U(n)} = SU(n) \times \mathbb{R}$, so the only new groups we can genera as universal covering groups are

$$\widetilde{SO(n)} \ .$$

We see that $\rho : \widetilde{SO(n)} \to SO(n)$ is a 2-1 homomorphism. We construct and study these new groups in the next chapter.

E. <u>Exercises</u>

1. Let S^1 be the additive group of reals mod 1 . Show that if $x \in [0,1]$ is irrational, then x is a generator for S^1 . (Hint: For $y = \frac{p}{q}$ rational, $\{ny \mid n = 1,2,\ldots\}$ gets within $\frac{1}{q}$ of each point in S^1 . Show that if x is irrational we can write $x = \lim_{i\to\infty} y_i$ where each $y_i = \frac{p_i}{q_i}$ is rational and $\lim_{i\to\infty} q_i = \infty$.)

2. Let G be a matrix group which is not necessarily connected.

Let $G^\circ = \left\{ x \in G \mid \exists \text{ path } \omega : [0,1] \to G \atop \omega(0) = 0 \ \omega(1) = x \right.$.

Call this the <u>identity component</u> of G . Show that G° is a subgroup of G .

3. Let G be a matrix group and H a normal subgroup of G. Show that the identity component H° of H is also a normal subgroup of G .

Chapter 10
Spin(k)

A. Clifford Algebras

One way of constructing groups which are subsets of some $\mathbb{R}^n$ is: Let $\mathcal{G}$ be a finite-dimensional real algebra and let G be the group of units in $\mathcal{G}$. We get more groups as subgroups of G. For example, we have used the algebra $M_n(\mathbb{R})$ in which the group of units is $GL(n,\mathbb{R})$ and we have the important subgroup $SO(n)$. Our groups $Spin(k)$ are subgroups of the group of units in the Clifford algebra C_k.

For $k = 0,1,2,\ldots$ we will define a real algebra C_k of dimension 2^k. First we set

$$C_0 = \mathbb{R} .$$

Next, C_1 is to be two dimensional: we take a basis $1,e$; let 1 act as a multiplicative identity and define $e^2 = -1$. This specifies the multiplication

$$(a + be)(c + de) = (ac - bd) + (ad + bc)e$$

and we see that

$$C_1 \cong \mathbb{C}$$

as 2-dimensional real algebras.

Next C_2 is to be 4-dimensional. We take a basis

$$1 , e_1 , e_2 , e_1e_2$$

and set $e_1^2 = -1$, $e_2^2 = -1$ and $e_2e_1 = -e_1e_2$. This specifies the multiplication and we see that the assignment

$$\begin{aligned} 1 &\mapsto 1 \\ e_1 &\mapsto i \\ e_2 &\mapsto j \\ e_1e_2 &\mapsto k \end{aligned}$$

gives an isomorphism

$$C_2 \cong \mathbb{H} .$$

<u>Definition</u>: C_k is the algebra which is generated, as an algebra, by $e_1,e_2,\ldots,e_k$ subject to

$$e_i^2 = -1 \text{ and } e_je_i = -e_ie_j \text{ if } i \neq j .$$

To explain what is meant by "generated as an algebra," consider C_3 . To e_1,e_2,e_3 we must add 1 and products of the e_i's to get a vector-space basis for C_3 , and we get

$$\{1,e_1,e_2,e_3,e_1e_2,e_1e_3,e_2e_3,e_1e_2e_3\}$$

which gives dimension $2^3 = 8$. The same argument shows that

$$\{e_{i_1} \ldots e_{i_r} \mid i_1 < i_2 <\ldots< i_r , 0 \leq r \leq k\}$$

is a vector-space basis for C_k (r = 0 denotes the element 1 of the basis).

Note that sending e_i in C_{k-1} to e_i in C_k imbeds C_{k-1} as a subalgebra of C_k.

$$C_0 \subset C_1 \subset \cdots \subset C_{k-1} \subset C_k \subset \cdots .$$

Now C_0, C_1 are fields and C_2 is a skew field but we cannot expect this for $k > 2$.

__Proposition__ 1: _For all_ $k > 2$, C_k _has divisors of zero._

__Proof__: It suffices to find a divisor of zero in C_3; i.e., an $x \neq 0$ such that there exists a $y \neq 0$ with $xy = 0$. Well,

$$(1 + e_1e_2e_3)(1 - e_1e_2e_3) = 1 - e_1e_2e_3e_1e_2e_3 = 0 .$$

There are two notable ways of getting algebras which have divisors of zero.

__Definition__: If $\mathfrak{A}$ and $\mathfrak{B}$ are real algebras, their _direct sum_ $\mathfrak{A} \oplus \mathfrak{B}$ is the set of all pairs (x,y) with $x \in \mathfrak{A}$ and $y \in \mathfrak{B}$ with operations

$$(x,y) + (z,w) = (x+z, y+w)$$

$$r(x,y) = (rx, ry)$$

$$(x,y)(z,w) = (xz, yw) .$$

__Proposition__ 2: $\mathfrak{A} \oplus \mathfrak{B}$ _has divisors of zero, and for any field_ F, $M_n(F)$ _has divisors of zero if_ $n \geq 2$.

<u>Proof</u>:

$$(1,0)(0,1) = (0.0)$$

$$\begin{pmatrix} 1 & 1 \\ 1 & 1 \end{pmatrix}\begin{pmatrix} 1 & -1 \\ -1 & 1 \end{pmatrix} = \begin{pmatrix} 0 & 0 \\ 0 & 0 \end{pmatrix} .$$

For our purposes we need to know the Clifford algebras up through C_5. There is a uniform way of getting up through C_4 (due to Alan Wiederhold). We consider

$$\mathbb{C} \subset \mathbb{H} \subset \mathbb{H} \oplus \mathbb{H} \subset M_2(\mathbb{H})$$

as follows: $\alpha \in \mathbb{C}$ is represented by $\begin{pmatrix} \alpha & 0 \\ 0 & \alpha \end{pmatrix}$, $q \in \mathbb{H}$ by $\begin{pmatrix} q & 0 \\ 0 & q \end{pmatrix}$ and $(q_1, q_2) \in \mathbb{H} \oplus \mathbb{H}$ by $\begin{pmatrix} q_1 & 0 \\ 0 & q_2 \end{pmatrix}$. Then the assignment

$$\begin{pmatrix} i & 0 \\ 0 & i \end{pmatrix} \to e_1$$

$$\begin{pmatrix} j & 0 \\ 0 & j \end{pmatrix} \to e_2$$

$$\begin{pmatrix} k & 0 \\ 0 & -k \end{pmatrix} \to e_3$$

$$\begin{pmatrix} 0 & k \\ k & 0 \end{pmatrix} \to e_4$$

gives isomorphisms

$$\mathbb{C} \xrightarrow{\approx} C_1$$

$$\mathbb{H} \xrightarrow{\approx} C_2$$

$$\mathbb{H} \oplus \mathbb{H} \xrightarrow{\approx} C_3$$

$$M_2(\mathbb{H}) \xrightarrow{\approx} C_4 .$$

Finally, the assignment

$$\begin{pmatrix} i & 0 & 0 & 0 \\ 0 & -i & 0 & 0 \\ 0 & 0 & i & 0 \\ 0 & 0 & 0 & -i \end{pmatrix} \quad \begin{pmatrix} 0 & -1 & 0 & 0 \\ 1 & 0 & 0 & 0 \\ 0 & 0 & 0 & -1 \\ 0 & 0 & 1 & 0 \end{pmatrix} \quad \begin{pmatrix} 0 & i & 0 & 0 \\ i & 0 & 0 & 0 \\ 0 & 0 & 0 & -i \\ 0 & 0 & -i & 0 \end{pmatrix}$$

$$\downarrow \qquad\qquad \downarrow \qquad\qquad \downarrow$$

$$e_1 \qquad\qquad e_2 \qquad\qquad e_3$$

$$\begin{pmatrix} 0 & 0 & 0 & -1 \\ 0 & 0 & -1 & 0 \\ 0 & 1 & 0 & 0 \\ 1 & 0 & 0 & 0 \end{pmatrix} \quad \begin{pmatrix} 0 & 0 & 0 & -i \\ 0 & 0 & -i & 0 \\ 0 & -i & 0 & 0 \\ -i & 0 & 0 & 0 \end{pmatrix}$$

$$\downarrow \qquad\qquad \downarrow$$

$$e_4 \qquad\qquad e_5$$

induces an isomorphism $M_4(\mathbb{C}) \xrightarrow{\cong} C_5$.

B. Pin(k) and Spin(k)

Let R^k denote the k-space in C_k spanned by $e_1,\ldots,e_k$ and let S^{k-1} be the unit sphere in R^k .

Proposition 3: *If C^*_k denotes the group of units in C_k , then $S^{k-1} \subset C^*_k$; i.e., each $x \in S^{k-1}$ is a unit.*

Proof: Let $x = a_1e_1 + \ldots + a_ke_k$ with $a_1^2 + \ldots + a_k^2 = 1$. Then

$$(a_1e_1 +\ldots+ a_ke_k)((-a_1)e_1 +\ldots+ (-a_k)e_k) = a_1^2 +\ldots+ a_k^2 = 1 .$$

Definition: Pin(k) is the subgroup of C^*_k generated by S^{k-1} .

Thus each element of Pin(k) is a finite product of elements of S^{k-1}.

We define a <u>conjugation</u> in C_k. It suffices to define it on basis elements and we set

$$(e_{i1}\cdots e_{ir})^* = (-e_{ir})\cdots(-e_{i1}) = (-1)^r e_{ir}\cdots e_{i1} .$$

For example,

$$1^* = 1 , \ e_i^* = -e_i , \ (e_ie_j)^* = -e_ie_j ,$$

$$(e_ie_je_k)^* = e_ie_je_k , \text{ etc.}$$

Clearly $(x^*)^* = x$. Also, we have $(xy)^* = y^*x^*$ because

$$((e_{i1}\cdots e_{ir})(e_{j1}\cdots e_{js}))^* = (-e_{js})\cdots(-e_{j1})(-e_{ir})\cdots(-e_{i1})$$

$$= (e_{j1}\cdots e_{js})^*(e_{i1}\cdots e_{ir})^* .$$

We also define an <u>automorphism</u> α of C_k by $\alpha(e_i) = -e_i$. (Note that conjugation is not an automorphism, because it agrees with α on each e_i, but $\alpha(e_ie_j) = (-e_i)(-e_j) = e_ie_j$ whereas $(e_ie_j)^* = -e_ie_j$.)

<u>Definition</u>: For $u \in \text{Pin}(k)$ and $x \in R^k$ we set

$$\rho(u)(x) = \alpha(u)\, x\, u^*.$$

It is not clear that $\rho(u)(x) \in R^k$ but it is a consequence of the next two propositions.

<u>Proposition</u> 4: <u>If</u> $u \in S^{k-1} \subset \text{Pin}(k)$ and $u \neq \pm 1$, <u>then</u> $\rho(u)$ <u>is reflection in</u> R^k <u>in the hyperplane perpendicular to</u> u.

Proof: Pick an orthonormal basis $u_1,\ldots,u_k$ of R^k with $u_1 = u$. Consider

$$\rho(u_1)u_i = \alpha(u_1)u_iu_1^* = (-u_1)u_i(-u_1) = u_1u_iu_1 \ .$$

If $i \neq 1$ this equals u_i and if $i = 1$ it equals $-u_1$, proving the proposition.

Proposition 5: ρ is a homomorphism of $\mathrm{Pin}(k)$ onto $\mathfrak{O}(k)$ and $\ker \rho = \{1,-1\}$.

Proof: Every element of $\mathrm{Pin}(k)$ is a finite product of elements of S^{k-1}, so to prove ρ is a homomorphism it suffices to take $u,v \in S^{k-1}$ and $x \in R^k$. Then

$$\rho(uv)(x) = \alpha(uv)x(uv)^* = \alpha(u)(\alpha(v)xv^*)u^* = \rho(u)(\rho(v)x) \ .$$

It now follows that for each $u \in \mathrm{Pin}(k)$, $\rho(u)$ does map R^k into R^k and it is orthogonal since it is a product of reflections. In Chapter VIII we showed that $\mathfrak{O}(k)$ is generated by reflections so that ρ is surjective.

It is clear that 1 and -1 are in the kernel of ρ. If we have $e_{i1}\cdots e_{ir}$ in the kernel of ρ with $r > 1$ (we cannot have $r = 1$) we get a contradiction as follows: If $\rho(e_{i1}\cdots e_{ir})$ is the identity, then for every $x \in R^k$ we must have

$$\alpha(e_{i1}\cdots e_{ir})x(e_{i1}\cdots e_{ir})^* = x \quad \text{or}$$

$$e_{i1}\cdots e_{ir}x = xe_{i1}\cdots e_{ir} \ .$$

Taking $x = e_{i1}$ gives $(-1)^{r-1}e_{i1}e_{i1}\cdots e_{ir} = e_{i1}e_{i1}\cdots e_{ir}$ and this gives $e_{i2} \cdots e_{ir} = 0$ since r must be even (the product of an odd number of reflections cannot be the identity).

<u>Definition</u>: $\mathrm{Spin}(k) = \rho^{-1}(SO(k))$.

Let us calculate $\mathrm{Spin}(1)$, $\mathrm{Spin}(2)$ and $\mathrm{Spin}(3)$. $C_1 = \mathbb{C}$ so $C_1^* = \mathbb{C} - (0)$ and $\mathrm{Pin}(1)$ is the subgroup of C_1^* generated by $S^0 = \{e_1, -e_1\}$. This is just $\{e_1, e_1^2 = -1, e_1^3 = -e_1, e_1^4 = 1\} \cong \mathbb{Z}/4$. Now $\rho(e_1) = \rho(-e_1)$ is a reflection and

$$\mathrm{Spin}(1) = \{1, -1\} \cong \mathbb{Z}/2 \quad .$$

$C_2 = \mathbb{H}$ so $C_2^* = \mathbb{H} - (0)$ and $\mathrm{Pin}(2)$ is the subgroup generated by

$$S^1 = \{ae_1 + be_2 \mid a^2 + b^2 = 1\} \quad .$$

Now $(ae_1 + be_2)(ce_1 + de_2) = -(ac + bd) + (ad - bc)e_1e_2$, so that $\mathrm{Pin}(2)$ must contain

$$S = \{c + de_1e_2 \mid c^2 + d^2 = 1\} \, .$$

Actually it is easy to see that

$$\mathrm{Pin}(2) = S \cup S^1 \text{ and } \mathrm{Spin}(2) = S \quad .$$

Similar computations easily give

$$\mathrm{Spin}(3) = \{a + be_1e_2 + ce_1e_3 + de_2e_3 \mid a^2 + b^2 + c^2 + d^2 = 1\}$$

and we see that

$$\mathrm{Spin}(3) \cong Sp(1) \; .$$

We conclude this section by calculating the centers of the

spin groups.

First off we always have $\{1,-1\}$ in the center. This is all of the center in Spin(1) (being all of Spin(1)) and Spin(2) equals its center. So we now assume

$$k \geq 3 .$$

Suppose $e_{i1}\cdots e_{ir}$ fails to contain e_j . Consider

$$(e_{i1}\cdots e_{ir})(e_{ir}e_j) = (-1)^{r-1}e_{ir}(e_{i1}\cdots e_{ir})e_j$$

$$= (-1)^{r-1}(-1)^r(e_{ir}e_{ij})(e_{i1}\cdots e_{ir})$$

$$= -(e_{ir}e_{ij})(e_{i1}\cdots e_{ir}) ,$$

proving $e_{i1}\cdots e_{ir}$ is not in the center. Thus $e_1e_2\cdots e_k$ and $-e_1e_2\cdots e_k$ are the only candidates (besides 1 and -1). Now for k odd $e_1\cdots e_k$ is not in Spin(k) . For $k = 2n$ we see that

$$(e_1e_2\cdots e_{2n})(e_{i1}\cdots e_{ir}) \quad (\text{ r even })$$

$$= (e_{i1}\cdots e_{ir})(e_1e_2\cdots e_{2n})(-1)^{(2n-1)r} = (e_{i1}\cdots e_{ir})(e_1e_2\cdots e_n)$$

so that $e_1\cdots e_{2n}$ is in the center.

__Proposition__ 6: (i) __If__ k __is odd__ Center Spin(k) = $\{1,-1\}$;
(ii) Center Spin(2n) = $\mathbb{Z}/4$ __if__ n __is odd__;
(iii) Center Spin(2n) = $\mathbb{Z}/2 \oplus \mathbb{Z}/2$ __if__ n __is even__.

__Proof__: We have proved (i) and we have seen that Center Spin(2n) = $\{1,-1,e_1\cdots e_{2n},-e_1\cdots e_{2n}\}$. Note that $(e_1\cdots e_{2n})(e_1\cdots e_{2n}) = (-1)^{n(2n+1)}$ which is 1 if n is even and is -1 if n is odd. This proves the proposition.

C. The isomorphisms

We have already noted that if $e_{i1}\cdots e_{ir} \in \mathrm{Spin}(k)$ then r is even. In this case we have

$$(e_{i1}\cdots e_{ir})(e_{i1}\cdots e_{ir})^{*} = 1 .$$

So that if $x \in \mathrm{Spin}(k)$, x is a linear combination of "even-graded" basis elements and $xx^{*} = 1$.

Definition: Define $\phi : C_{k-1} \to C_k$ by

$$\phi(e_i) = e_i e_k \quad \text{for } i = 1,\dots,k-1 .$$

Then ϕ extends to an algebra homomorphism since $(e_i e_k)^2 = -1$ and $(e_i e_k)(e_j e_k) = -(e_j e_k)(e_i e_k)$. Indeed, $e_i e_k e_j e_k = e_i e_j$ and we see that ϕ is an isomorphism of C_{k-1} onto the even-graded part of C_k .

Proposition 7: The isomorphisms ψ (§A) and ϕ satisfy:

$$(\phi(x))^{*} = \phi(x^{*})$$

$$\psi({}^{t}\bar{M}) = (\psi(M))^{*} .$$

Proof: $(\phi(e_i))^{*} = (e_i e_k)^{*} = (-e_k)(-e_i) = -e_i e_k = \phi(e_i^{*})$.

For C_1, C_2, C_3, C_4 we check the second formula on the basic matrices

$$M = \begin{pmatrix} i & 0 \\ 0 & i \end{pmatrix} \overset{\psi}{\to} e_1 \qquad {}^t\bar{M} = \begin{pmatrix} -i & 0 \\ 0 & -i \end{pmatrix} \overset{\psi}{\to} -e_1 = (\psi(M))^* .$$

The others work the same way, and the verification for $M_4(\mathbb{C}) \overset{\psi}{\to} C_5$ is equally routine.

Theorem: The composed maps

$$Sp(2) \to M_2(\mathbb{H}) \overset{\psi}{\underset{\cong}{\to}} C_4 \overset{\phi}{\to} C_5$$

$$SU(4) \to M_4(\mathbb{C}) \overset{\psi}{\underset{\cong}{\to}} C_5 \overset{\phi}{\to} C_6$$

give isomorphism of $Sp(2)$ with $Spin(5)$ and $SU(4)$ with $Spin(6)$.

Proof: $Sp(2)$ and $SU(4)$ are those matrices such that $M\,{}^t\bar{M} = I$ so that

$$1 = \psi(M\,{}^t\bar{M}) = \psi(M)\psi(M)^* \quad \text{and}$$

$$1 = \phi\psi(M)\phi(\psi(M)^*) = \phi\psi(M)(\phi\psi(M))^* .$$

We get all even graded units x with $xx^* = 1$ in this manner. Thus

$$Spin(5) \subset \phi\psi(Sp(2))$$

$$Spin(6) \subset \phi\psi(SU(4)) ,$$

and $\phi\psi(Sp(2))$ is a closed connected manifold of dimension 10 and $Spin(5)$ is a closed manifold of dimension 10, so $Spin(5) = \phi\psi(Sp(2))$. Similarly, the two 15-dimensional manifolds $Spin(6)$ and $\phi\psi(SU(4))$ are equal.

D. Exercises

1. Show that

$$U(1) \to \mathbb{C} \xrightarrow[\cong]{\psi} C_1 \xrightarrow{\phi} C_2$$

gives an isomorphism $U(1) \cong \mathrm{Spin}(2)$.

2. Show that

$$Sp(1) \to \mathbb{H} \xrightarrow[\cong]{\psi} C_2 \xrightarrow{\phi} C_3$$

gives an isomorphism $Sp(1) \cong \mathrm{Spin}(3)$.

3. Show that

$$Sp(1) \times Sp(1) \to \mathbb{H} \oplus \mathbb{H} \xrightarrow[\cong]{\psi} C_3 \xrightarrow{\phi} C_4$$

gives an isomorphism $Sp(1) \times Sp(1) \cong \mathrm{Spin}(4)$.

4. Let G be an abelian group with exactly four elements. Show that $G \cong \mathbb{Z}/4$ if and only if G has an element of order 4 and $G \cong \mathbb{Z}/2 \oplus \mathbb{Z}/2$ otherwise.

Chapter 11
Normalizers, Weyl Groups

A. Normalizers

We saw in Chapter X that $Sp(1) \cong Spin(3)$ and $Sp(2) \cong Spin(5)$ (also that $Spin(6) \cong SU(4)$). The only remaining question in our isomorphism puzzle is whether $Sp(k)$ and $Spin(2k+1)$ are isomorphic for $k \geq 3$. In this chapter, we will use normalizers and Weyl groups to show that they are not.

Definition: Let S be a nonempty subset of a group G . Set

$$N(S) = \{x \in G \mid xSx^{-1} = S\}$$

and call $N(S)$ the normalizer (of S in G). $N(S)$ is a subgroup of G (Exercise 5, Chapter I).

Lemma 1: *If S is a subgroup of G , then S is a normal subgroup of $N(S)$.*

Proof: (Exercise 5. Chapter I).

We calculate now the normalizers of our standard maximal tori in S^3 ($= Sp(1)$) and $SO(3)$.

Let x be an element of a group G , and define

$$A_x : G \to G$$

by $A_x(g) = xgx^{-1}$ (conjugation by x). Then A_x is an isomorphism of G onto G.

Our standard maximal torus in $Sp(1)$ is $T = \{a+ib \mid a^2+b^2 = 1\}$. Let N be the normalizer of T in $Sp(1)$. If $q \in N$ then $A_q : T \to T$ is an isomorphism of the circle group T onto itself.

Lemma 2: There are exactly two isomorphisms of the circle group.

Proof: Let S^1 (as T above) be the set of complex numbers of unit length. Then $F = \{1,-1,i,-i\}$ is the set of all 4th roots of unity in S^1. If $\phi : S^1 \to S^1$ is an isomorphism, it must have $\phi(1) = 1$, and therefore $\phi(-1) = -1$. Thus $\phi(i) \in \{i,-i\}$. Both possibilities give isomorphisms and knowing ϕ on F determines ϕ. (See Exercise #5.) q.e.d.

Let $q \in N$. If A_q is the identity then q commutes with each $x \in T$, so $q \in T$. So suppose A_q is the other isomorphism. Since $|q| = 1$, $q^{-1} = \bar{q}$, so we have

$$qi\bar{q} = -i .$$

If we write $q = x + iy + jz + kw$, this gives

$$x^2 + y^2 - z^2 - w^2 = -1 ,$$

and since

$$x^2 + y^2 + z^2 + w^2 = 1$$

we get $x = 0 = y$. Furthermore, if $q = jz + kw$ with $z^2 + w^2 = 1$

we do get $A_q(i) = -i$. Hence

$$N = T \cup T^* \quad \text{(disjoint union)}$$

where $T^* = \{jz + kw \mid z^2 + w^2 = 1\}$. Thus the quotient group $N/T = \mathbb{Z}/2$ (T is normal in N), and we have

$$T \to N \to \mathbb{Z}/2 .$$

Next we calculate the normalizer N' of the standard maximal torus T' in $SO(3)$. Recall that

$$T' = \left\{ \begin{pmatrix} \cos\theta & \sin\theta & 0 \\ -\sin\theta & \cos\theta & 0 \\ 0 & 0 & 1 \end{pmatrix} \right\} .$$

Now $T' \subset N'$ and we seek the elements of $N' - T'$.

Lemma 3: $N' - T' = \left\{ \begin{pmatrix} \cos\phi & \sin\phi & 0 \\ \sin\phi & -\cos\phi & 0 \\ 0 & 0 & -1 \end{pmatrix} \right\}$.

Proof: Each such matrix is symmetric and is its own inverse and has determinant $+1$, so it is in $SO(3)$. It is a routine, but instructive, exercise to show that

$$\begin{pmatrix} \cos\phi & \sin\phi & 0 \\ \sin\phi & -\cos\phi & 0 \\ 0 & 0 & -1 \end{pmatrix} \begin{pmatrix} \cos\theta & \sin\theta & 0 \\ -\sin\theta & \cos\theta & 0 \\ 0 & 0 & 1 \end{pmatrix} \begin{pmatrix} \cos\phi & \sin\phi & 0 \\ \sin\phi & -\cos\phi & 0 \\ 0 & 0 & -1 \end{pmatrix} = \begin{pmatrix} \cos\theta & -\sin\theta & 0 \\ \sin\theta & \cos\theta & 0 \\ 0 & 0 & 1 \end{pmatrix} .$$

It is less routine (but also instructive) to show that these are the only matrices in $SO(3)$ which give the nontrivial isomorphism of T' .

So we have:

$$T \to N \to \mathbb{Z}/2 \quad \text{for} \quad Sp(1)$$

$$T' \to N' \to \mathbb{Z}/2 \quad \text{for} \quad SO(3)$$

and both T and T' are circle groups.

Proposition 1: $N' \not\cong N$.

Proof: If $\phi : N \to N'$ were an isomorphism it would have to send T to T' and $N - T$ to $N' - T'$. If $x \in N - T$ we would have $\phi(x^2) = (\phi(x))^2$. But it is easy to check that if $x \in N - T$, then $x^2 = -1$ and if $y \in N' - T'$, then $y^2 = I$.

So we see that N and N' are groups having circle subgroups and isomorphic quotient groups (namely $\mathbb{Z}/2$), but N and N' are not isomorphic. They are different extensions of S^1 by $\mathbb{Z}/2$.

Definition: Let A be an abelian normal subgroup of B and C be the quotient group,

$$A \overset{\alpha}{\to} B \overset{\beta}{\to} C$$

α being the inclusion and β being the quotient map. We say say this extension splits if there is a homomorphism $\gamma : C \to B$ such that $\beta \circ \gamma$ is the identity on C . (γ is called a homomorphic cross section.)

Then $T' \to N' \to \mathbb{Z}/2$ splits (send identity to identity and send the non identity element of $\mathbb{Z}/2$ to $\begin{pmatrix} 1 & 0 & 0 \\ 0 & -1 & 0 \\ 0 & 0 & -1 \end{pmatrix}$), but $T \to N \to \mathbb{Z}/2$ does not split (because we must send identity to identity and the non identity element of $\mathbb{Z}/2$ into $N - T$ and the result is not a homomorphism).

This is the way we are going to distinguish $Sp(k)$ and $Spin(2k+1)$ for $k \geq 3$.

B. Weyl groups

Let T be a maximal torus in a matrix group G and N be the normalizer of T. We have

$$T \overset{\alpha}{\to} N \overset{\beta}{\to} W$$

where $W = N/T$, α is the inclusion and β is the quotient map.

Definition: W is called the Weyl group of G. We note that if we used another maximal torus T', then $T' = xTx^{-1}$ and $A_x : G \to G$ would induce an isomorphism of T to T', N to the normalizer N' of T' and hence induce an isomorphism of Weyl groups (Exercise). Similarly, isomorphic groups will have isomorphic Weyl groups.

Proposition 2: Let $x, y \in N$. Then A_x and A_y are the same isomorphism on $T \Leftrightarrow x$ and y belong to the same coset of T in N.

Proof: $\Leftarrow$

$x = y\tau$ for some $\tau \in T$. So $A_x(t) = xtx^{-1} = y\tau t(y\tau)^{-1} = y\tau t\tau^{-1}y^{-1}$, and since T is abelian this is just $yty^{-1} = A_y(t)$.

$\Rightarrow$

If $xtx^{-1} = yty^{-1}$ for all $t \in T$, then $(y^{-1}x)t(y^{-1}x)^{-1} = t$ so that $y^{-1}x$ commutes with all $t \in T$ and T is maximal so $y^{-1}x \in T$.

<u>Corollary</u>: W <u>acts on</u> T.

<u>Proof</u>: For $w \in W$ choose $x \in \beta^{-1}(w)$ and define $w(t) = A_x(t)$. By Proposition 2, this is independent of the choice of x.

<u>Proposition</u> 3: W <u>is a finite group</u>.

<u>Proof</u>: Since N is compact, W is compact. Since T is the identity component of N, W is discrete. Thus W is finite.

From this it follows that N is an extension of T by a finite group. In particular then, T is the connected component of N containing the identity. We have

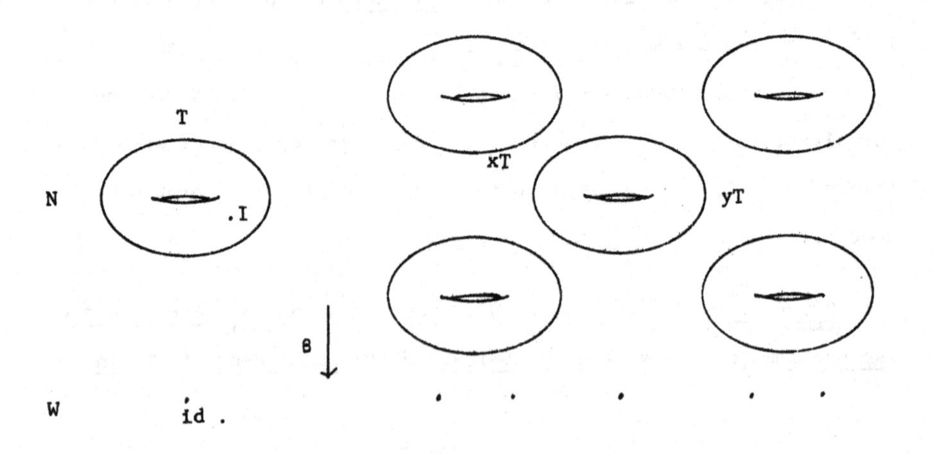

C. Spin(2n+1) and Sp(n)

An isomorphism of Spin(2n+1) onto Sp(n) would induce an isomorphism of their centers and would induce

$$\frac{\mathrm{Spin}(2n+1)}{\mathrm{center}} \cong \frac{\mathrm{Sp}(n)}{\mathrm{center}} .$$

Theorem: For $n = 3,4,5,\ldots$ the normalizer of a maximal torus in $\frac{\mathrm{Spin}(2n+1)}{\mathrm{center}}$ splits, but the normalizer of a maximal torus in $\frac{\mathrm{Sp}(n)}{\mathrm{center}}$ does not split. So for $n = 3,4,5,\ldots$

$$\mathrm{Spin}(2n+1) \not\cong \mathrm{Sp}(n) .$$

We will prove slightly more; namely,

(*) The normalizer in Sp(n) does not split for any n ;

(†) The normalizer in $\frac{\mathrm{Sp}(n)}{\mathrm{center}}$ splits $\Leftrightarrow n \in \{1,2\}$;

(T) The normalizer in $\frac{\mathrm{Spin}(2n+1)}{\mathrm{center}}$ (= SO(2n+1)) splits for $n = 1,2,3,\ldots$.

Proof of (*): We want to show that

$$T \xrightarrow{\alpha} N \underset{\overset{\dashleftarrow}{\gamma}}{\xrightarrow{\beta}} W$$

does not split in Sp(n) ; i.e. that no homomorphic cross-section γ can exist. We will suppose γ exists and obtain a contradiction.

Let $\eta = \begin{pmatrix} j_1 & & 0 \\ & \ddots & \\ 0 & & 1 \end{pmatrix} \in \mathrm{Sp}(n)$. We claim that $\eta \in N$. Take any

$t \in \begin{pmatrix} e^{i\theta_1} & & \bigcirc \\ & \ddots & \\ \bigcirc & & e^{i\theta_n} \end{pmatrix}$ in T and we calculate

$$\eta t \eta^{-1} = \begin{pmatrix} j_1 & & \bigcirc \\ & \ddots & \\ \bigcirc & & 1 \end{pmatrix} \begin{pmatrix} e^{i\theta_1} & & \bigcirc \\ & \ddots & \\ \bigcirc & & e^{i\theta_n} \end{pmatrix} \begin{pmatrix} -j_1 & & \bigcirc \\ & \ddots & \\ \bigcirc & & 1 \end{pmatrix}$$

$$= \begin{pmatrix} je^{i\theta_1}(-j) & & & \bigcirc \\ & e^{i\theta_2} & & \\ & & \ddots & \\ \bigcirc & & & e^{i\theta_n} \end{pmatrix}$$

and $je^{i\theta_1}(-j) = j(\cos\theta_1 + i\sin\theta_1)(-j) = e^{-i\theta_1}$, so that $\eta t \eta^{-1}$ is again in T.

Let $w = \beta(\eta)$. Then, since $\eta^2 = \begin{pmatrix} -1_1 & & \bigcirc \\ & \ddots & \\ \bigcirc & & 1 \end{pmatrix} \in T$ we have that $w^2 = \beta(\eta^2) = 1 \in W$. On the other hand, since γ is assumed to be a cross section we have

$$\eta' = \gamma(w)$$

must be in the same coset of T as η is, and since γ is to be a homomorphism we must have

$$(\eta')^2 = \gamma(w^2) = \gamma(1) = I \quad .$$

So we can complete our proof of (*) by showing that no such η' exists; i.e. no element in the coset of η square to the identity.

Any such η' can be written $\eta' = \eta t$

$$\eta' = \begin{pmatrix} j_1 & & \\ & \ddots & \\ & & 1 \end{pmatrix} \begin{pmatrix} e^{i\theta_1} & & \\ & \ddots & \\ & & e^{i\theta_n} \end{pmatrix}$$

$$\eta' = \begin{pmatrix} je^{i\theta_1} & & & 0 \\ & e^{i\theta_2} & & \\ & & \ddots & \\ 0 & & & e^{i\theta_n} \end{pmatrix}$$

Thus

$$(\eta')^2 = \begin{pmatrix} je^{i\theta_1}je^{i\theta_1} & & & 0 \\ & e^{2i\theta_2} & & \\ & & \ddots & \\ 0 & & & e^{2i\theta_n} \end{pmatrix}$$

$$= \begin{pmatrix} -1 & & & 0 \\ & e^{2i\theta_2} & & \\ & & \ddots & \\ 0 & & & e^{2i\theta_n} \end{pmatrix} \neq I$$

and (*) is proved.

Proof of (+): We want to prove that the normalizer in $\frac{Sp(n)}{center}$ splits $\Leftrightarrow n \in \{1,2\}$. For $n = 1$, $\frac{Sp(1)}{center} = SO(3)$ and we have seen that N splits for SO(3) (§A). For $n = 2$ we have $Sp(2) \cong Spin(5)$ and $\frac{Sp(2)}{center} \cong SO(5)$ and we will show in (T') that SO(odd) always splits. So it suffices to show here that $\frac{Sp(n)}{center}$

does not split for $n \geq 3$.

For notational convenience we will prove this for $n = 3$ and the general argument will be clear. The idea is like that used for the proof of (*). We need to find a relation in W which cannot lift (via γ) to a relation in N.

Let $\eta = \begin{pmatrix} j & 0 & 0 \\ 0 & 1 & 0 \\ 0 & 0 & 1 \end{pmatrix}$ and $\sigma = \begin{pmatrix} 0 & 1 & 0 \\ 1 & 0 & 0 \\ 0 & 0 & 1 \end{pmatrix}$ and one checks easily that η, σ are in N . Let

$$w_1 = \beta(\eta) \quad \text{and} \quad w_2 = \beta(\sigma) \quad .$$

Also since η^2 and σ^2 are in T we have

$$w_1^2 = 1 \quad \text{and} \quad w_2^2 = 1 \quad .$$

Also

$$\eta\sigma = \begin{pmatrix} 0 & j & 0 \\ 1 & 0 & 0 \\ 0 & 0 & 1 \end{pmatrix} \quad .$$

It is left as an exercise to show that the fourth power of $\eta\sigma$ lies in T . So we have

$$(w_1 w_2)^4 = 1 \quad .$$

If we had a homomorphic cross section then we would have to have

$$(\gamma(w_1))^2 \in \{I,-I\} \quad (= \text{center})$$
$$(\gamma(w_2))^2 \in \{I,-I\}$$
$$(\gamma(w_1)\gamma(w_2))^4 \in \{I,-I\} \quad .$$

So we will assume that we have η' in the coset of η, σ' in the coset of σ such that $(\eta')^2$, $(\sigma')^2$ and $(\eta'\sigma')^4$ are all in $\{I,-I\}$.

Lemma a: $(\eta')^2 = -I$.

Proof: We have

$$\eta' = \eta t = \begin{pmatrix} je^{i\theta_1} & 0 & 0 \\ 0 & e^{i\theta_2} & 0 \\ 0 & 0 & e^{i\theta_3} \end{pmatrix}$$

and $(\eta')^2$ cannot be I since it has -1 in the $1,1$ position.

Thus $(\eta')^2 = -I$ and $\eta' = \begin{pmatrix} je^{i\theta_1} & 0 & 0 \\ 0 & \pm i & 0 \\ 0 & 0 & \pm i \end{pmatrix}$.

Lemma b: For some positive integer m , σ' may be written as

$$\sigma' = \begin{pmatrix} 0 & (-1)^m e^{-i\phi} & 0 \\ e^{i\phi} & 0 & 0 \\ 0 & 0 & i^m \end{pmatrix} .$$

Proof: We have

$$\sigma' = \sigma t = \begin{pmatrix} 0 & 1 & 0 \\ 1 & 0 & 0 \\ 0 & 0 & 1 \end{pmatrix} \begin{pmatrix} e^{i\phi_1} & 0 & 0 \\ 0 & e^{i\phi_2} & 0 \\ 0 & 0 & e^{i\phi_3} \end{pmatrix}$$

$$\sigma' = \begin{pmatrix} 0 & e^{i\phi_2} & 0 \\ e^{i\phi_1} & 0 & 0 \\ 0 & 0 & e^{i\phi_3} \end{pmatrix} .$$

$$\text{Then} \quad (\sigma')^2 = \begin{pmatrix} e^{i(\phi_1+\phi_2)} & 0 & 0 \\ 0 & e^{i(\phi_1+\phi_2)} & 0 \\ 0 & 0 & e^{2i\phi_3} \end{pmatrix}$$

and this must be I or $-I$. This shows that $e^{2i\phi_3}$ is 1 or -1 so the $e^{i\phi_3} = i^m$ for some m. If m is odd $(\sigma')^2$ must be $-I$ and then $e^{i\phi_1}e^{i\phi_2} = -1$ shows we can take $\phi_2 = \pi - \phi_1$. If m is even we can take $\phi = \phi_1$ and $e^{i\phi_2} = (-1)^m e^{i\phi}$, and Lemma b is proved. We finish $(\dagger)$ by

<u>Lemma</u> c: $(\eta'\sigma')^4 = \begin{pmatrix} -1 & 0 & 0 \\ 0 & -1 & 0 \\ 0 & 0 & 1 \end{pmatrix} \notin \{I, -I\}$.

<u>Proof</u>: From Lemmas a and b we have

$$\eta'\sigma' = \begin{pmatrix} 0 & je^{i\theta}(-1)^m e^{i\phi} & 0 \\ \pm ie^{-i\phi} & 0 & 0 \\ 0 & 0 & (\pm 1)i^m \end{pmatrix}$$

and a routine (if somewhat tedious) calculation proves the lemma and hence $(\dagger)$.

D. SO(n) <u>splits</u>

We showed that the normalizer in SO(3) split, but that was easy because we knew the Weyl group was just $\mathbb{Z}/2$, so we simply had to find an element of $N - T$ whose square equalled I . (It turned out that any element of $N - T$ would do.) For $n > 3$ we will have bigger Weyl groups. For notational simplicity we will work with $SO(5)$ first and then consider the general case.

Let

$$\tau(\theta_1,\theta_2) = \begin{pmatrix} \cos\theta_1 & \sin\theta_1 & 0 & 0 & 0 \\ -\sin\theta_1 & \cos\theta_1 & 0 & 0 & 0 \\ 0 & 0 & \cos\theta_2 & \sin\theta_2 & 0 \\ 0 & 0 & -\sin\theta_2 & \cos\theta_2 & 0 \\ 0 & 0 & 0 & 0 & 1 \end{pmatrix}$$

be an element in our maximal torus. Let P_1, P_2 be the two 2-planes in $\mathbb{R}^5$ spanned by coordinates 1,2 and 3,4 respectively. Let $\mathbb{R}$ denote the line of the 5th coordinate.

Proposition 4: *$P_1, P_2, \mathbb{R}$ are the only minimal subspaces stable for each $\tau(\theta_1,\theta_2)$ in T. If N is the normalizer of T and $A \in N$, then*

$$\mathbb{R}A = \mathbb{R} \text{ and}$$

$$P_1A \in \{P_1,P_2\} , \; P_2A \in \{P_1,P_2\} .$$

Proof: The first statement is clear. Now $\mathbb{R}A$ is some line, say $\mathbb{R}A = \mathbb{R}'$. Since $A \in N$ for every $\tau \in T$ we have $A\tau A^{-1} \in T$, so that $\mathbb{R}A\tau A^{-1} = \mathbb{R}$. Then

$$\mathbb{R}'\tau A^{-1} = \mathbb{R} \quad \text{or} \quad \mathbb{R}'\tau = \mathbb{R}A = \mathbb{R}' ,$$

proving that $\mathbb{R}' = \mathbb{R}$.

Now P_1A is some 2-plane, say $P_1A = P$. For any $\tau \in T$ we have

$$P_1 A \tau A^{-1} = P_1 \quad .$$

So $P\tau = P_1 A = P$. Thus $P \in \{P_1, P_2\}$. Of course the same proof works for P_2 .

Consider two specific elements of N .

$$A = \begin{pmatrix} 1 & 0 & 0 & 0 & 0 \\ 0 & 1 & 0 & 0 & 0 \\ 0 & 0 & 0 & 1 & 0 \\ 0 & 0 & 1 & 0 & 0 \\ 0 & 0 & 0 & 0 & -1 \end{pmatrix} \qquad B = \begin{pmatrix} 0 & 0 & 0 & 1 & 0 \\ 0 & 0 & 1 & 0 & 0 \\ 0 & 1 & 0 & 0 & 0 \\ 1 & 0 & 0 & 0 & 0 \\ 0 & 0 & 0 & 0 & 1 \end{pmatrix} \quad .$$

We have $A^{-1} = A$ and we easily find that

$$A\tau(\theta_1, \theta_2)A = \tau(\theta_1, -\theta_2) \quad .$$

Also $B = B^{-1}$ and

$$B\tau(\theta_1, \theta_2)B = \tau(-\theta_2, -\theta_1) \quad .$$

If we think of these as actions on the tangent space $\mathfrak{L}(T)$ of T (i.e., take differentials) we see that these are reflections

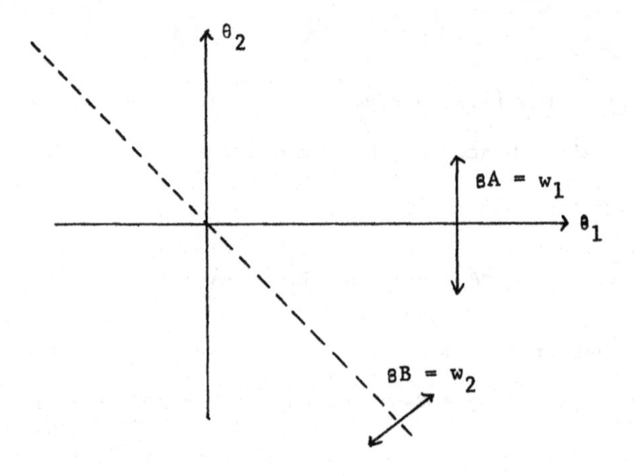

We have $T \overset{\alpha}{\to} N \overset{\beta}{\to} W$. Let $w_1 = \beta(A)$ and $w_2 = \beta(B)$. It is not hard to see that w_1 and w_2 generate a subgroup W' of W with W' containing all permutations and sign changes of $\{\theta_1, \theta_2\}$.

Writing $\mathbb{R}^5 = P_1 \oplus P_2 \oplus \mathbb{R}$ and using Proposition A we can easily limit the possibilities for $A \in N \subset SO(5)$. A induces orthogonal on P_1 and on P_2 or switching P_1 and P_2. Also $\det A = +1$. It turns out that there are only eight possibilities. Thus W is generated by W_1, W_2 with $W_1^2 = 1$, $W_2^2 = 1$ and $(W_1 W_2)^4 = 1$.

We have now that $SO(5)$ splits. For if we let $\gamma(1) = I$, $\gamma(w_1) = A$, $\gamma(w_2) = B$ we see that it only remains to check that

$$(AB)^4 = I .$$

Well,

$$AB = \begin{pmatrix} 0 & 0 & 0 & 1 & 0 \\ 0 & 0 & 1 & 0 & 0 \\ 1 & 0 & 0 & 0 & 0 \\ 0 & 1 & 0 & 0 & 0 \\ 0 & 0 & 0 & 0 & -1 \end{pmatrix} .$$

Then

$$(AB)^2 = \begin{pmatrix} 0 & 1 & 0 & 0 & 0 \\ 1 & 0 & 0 & 0 & 0 \\ 0 & 0 & 0 & 1 & 0 \\ 0 & 0 & 1 & 0 & 0 \\ 0 & 0 & 0 & 0 & 1 \end{pmatrix}$$

and so

$$(AB)^4 = \begin{pmatrix} 1 & 0 & 0 & 0 & 0 \\ 0 & 1 & 0 & 0 & 0 \\ 0 & 0 & 1 & 0 & 0 \\ 0 & 0 & 0 & 1 & 0 \\ 0 & 0 & 0 & 0 & 1 \end{pmatrix} = I .$$

We have noticed before that the groups SO(odd) and SO(even) are somewhat different. Let us analyze SO(4) just as we have SO(5). Let $C = \begin{pmatrix} 0 & 1 & 0 & 0 \\ 1 & 0 & 0 & 0 \\ 0 & 0 & 0 & 1 \\ 0 & 0 & 1 & 0 \end{pmatrix}$. Then $C^{-1} = C$ and

$$C\tau(\theta_1,\theta_2)C = \tau(-\theta_1,-\theta_2) \quad .$$

Let $B = \begin{pmatrix} 0 & 0 & 0 & 1 \\ 0 & 0 & 1 & 0 \\ 0 & 1 & 0 & 0 \\ 1 & 0 & 0 & 0 \end{pmatrix}$. Then $B \in SO(4)$, $B^{-1} = B$ and

$$B\tau(\theta_1,\theta_2)B = \tau(-\theta_2,-\theta_1) \quad .$$

So $BC\tau(\theta_1,\theta_2)CB = \tau(\theta_2,\theta_1)$. So we can permute θ_1 and θ_2 and we can <u>change</u> <u>both</u> <u>signs</u>. We claim that no element of W can change just one sign. Any $D \in N$ must be of the form

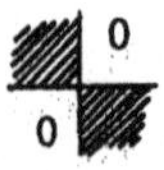

or of the form

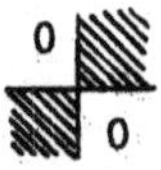

(see Proposition 4). And each shaded area is an orthogonal 2×2 matrix. If we had a $D \in N$ such that $D\tau(\theta_1,\theta_2)\,{}^tD$ changed just one sign but also permuted θ_1,θ_2 we could combine with B to get one changing one sign but not permuting. Such a matrix cannot have the form

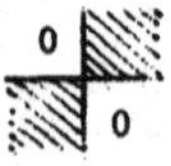

. But if $D = \begin{pmatrix} \sigma & 0 \\ 0 & \rho \end{pmatrix}$ with σ,ρ being 2×2 orthogonal matrices and $D\tau(\theta_1,\theta_2)\,{}^tD = \tau(-\theta_1,\theta_2)$ we easily calculate

that $\rho R(\theta_2)^t\rho = R(\theta_2)$ $(R(\theta) = \begin{pmatrix} \cos\theta & \sin\theta \\ -\sin\theta & \cos\theta \end{pmatrix})$. This implies that ρ commutes with all rotations. Thus ρ is a rotation, so $\det \rho = +1$. But then σ is a rotation and we cannot have

$$\sigma R(\theta_1)^t\sigma = R(-\theta_1)$$

as required.

Otherwise the homomorphic cross section proof goes just as for $SO(5)$. The elements $w_1 = \beta(C)$ and $w_2 = \beta(B)$ generate W and $\gamma(w_1) = C$, $\gamma(w_2) = B$ generates a homomorphic cross section. The general case is more difficult only because of notation. The Weyl group $SO(n)$ is generated by all permutations of $\{\theta_1,\ldots,\theta_n\}$ and all <u>even</u> sign changes.

E. Exercises

1. Show that an isomorphism of Lie groups induces an isomorphism of their Weyl groups.

2. Prove (+) for $\frac{Sp(n)}{\text{center}}$ ($n \geq 3$) (i.e., generalize the proof given for $\frac{Sp(3)}{\text{center}}$) .

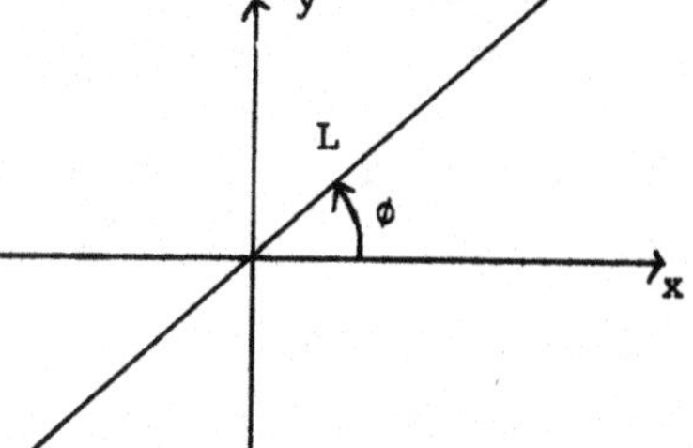

3. Show that reflection in L is given by

$$r = \begin{pmatrix} \cos 2\phi & \sin 2\phi \\ \sin 2\phi & -\cos 2\phi \end{pmatrix} .$$

4. Show that, with r as in 3 ,

$$r\begin{pmatrix}\cos\theta & \sin\theta\\ -\sin\theta & \cos\theta\end{pmatrix}r=\begin{pmatrix}\cos\theta & -\sin\theta\\ \sin\theta & \cos\theta\end{pmatrix} .$$

5. Show that the only isomorphism of $S^1 = \{e^{i\theta}\}$ which leaves i fixed is the identity. Show that there is exactly one isomorphism of S^1 which sends i to $-i$.

Chapter 12
Lie Groups

A. Differentiable manifolds

If U is an open set in $\mathbb{R}^n$ and we have a function $f : U \to \mathbb{R}^m$, we say that f is smooth (or C^∞) if all higher partial derivatives of f exist and are continuous. The composition of smooth functions is smooth. In the case $m = n$, if $f : U \to \mathbb{R}^n$ is one-to-one onto $f(U)$, with $f(U)$ open in $\mathbb{R}^n$, and both f and f^{-1} are smooth then f is a diffeomorphism (from U to $f(U)$).

Let M be a n-manifold (§E, Chapter VI). By definition, for any $p \in M$ we have U open in $\mathbb{R}^n$ and a homeomorphism

$$\phi : U \to M$$

into M with $\phi(U)$ being an open neighborhood of p. Such a pair (U,ϕ) will be called a chart. A collection of charts such that the $\phi(U)$'s cover M will be called an atlas.

Two charts $\phi : U \to M$ and $\psi : V \to M$ are said to overlap smoothly if (either $\phi(U) \cap \phi(V) = \emptyset$, or) $\psi^{-1} \circ \phi$ is a diffeomorphism.

Definition: M is a differentiable manifold if it has an atlas of smoothly overlapping charts.

Given such an atlas we can maximalize it. We simply add in all charts which overlap smoothly with the given ones in the atlas. Since compositions of diffeomorphisms are again diffeomorphisms we see that: If two new charts overlap smoothly with the given charts in the atlas, then they overlap smoothly with each other. We call a maximal atlas a differentiable structure.

Suppose we have differentiable manifolds M and N and a function $f : U \to N$ with U an open set in M . What should it mean for f to be smooth? For $x \in U$ choose charts $\phi : V \to M$ with $\phi(V)$ an open neighborhood of x and $\psi : W \to N$ with $\psi(W)$ an open neighborhood of $f(x)$. We say f is smooth at x if $\psi^{-1} \circ f \circ \phi$ is smooth at $\phi^{-1}(x)$. It is easy to check that this definition is independent of the choices of charts (within the maximal atlases).

A smooth one-to-one map $f : M \to N$ with a smooth inverse is a diffeomorphism of the differentiable manifolds. If $M = \mathbb{R} = N$ and f is given by $f(x) = x^3$, then f is smooth and one-to-one, but f^{-1} is not smooth (since its derivative fails to exist at 0) so that f is not a diffeomorphism.

B. Tangent vectors, vector fields

If we have a smooth curve or surface in some $\mathbb{R}^n$, then the concept of a vector tangent to the curve or surface is not hard to define. But our differentiable manifolds are to be thought of as spaces in their own right - not necessarily sitting in some euclidean space. How then do we define "tangent vector?" The idea is that if we have a vector in $\mathbb{R}^n$ we can use it to differentiate functions - essentially taking directional derivatives. So we will call a tangent vector a thing which differentiates functions.

Let M be a differentiable m-manifold and $p \in M$. We set

$$A(p) = \{(U,f) \mid p \in U \text{ open, } f : U \to \mathbb{R} \text{ smooth}\} .$$

We used $A(p)$ to denote this set of pairs (U,f) since we are going to make it into an algebra. It is routine to verify that if we define

$$(U,f) + (V,g) = (U \cap V, f+g) \quad \text{and}$$

$$r(U,f) = (U,rf) \quad ,$$

that these operations make $A(p)$ into a real vector space. Then we define

$$(U,f)(V,g) = (U \cap V, fg)$$

and $A(p)$ becomes an algebra.

<u>Definition</u>: A <u>tangent vector</u> η to M at p is a linear map $\eta : A(p) \to \mathbb{R}$ such that:

(i) if $f = g$ on some neighborhood of p, then $\eta(f) = \eta(g)$.
(ii) $\eta(fg) = f(p)\eta(g) + \eta(f)g(p)$.

Suppose f is a constant function, $f(x) = r$ for all x in some neighborhood of p. Then $fg = rg$ in that neighborhood, and if η is any tangent vector we have $\eta(rg) = r\eta(g)$, since η is linear. On the other hand, by (ii) we have

$$\begin{aligned}\eta(rg) = \eta(fg) &= f(p)\eta(g) + \eta(f)g(p) \\ &= r\eta(g) + \eta(f)g(p) \quad .\end{aligned}$$

So we see that $\eta(f)g(p) = 0$ for all g. Thus $\eta(f) = 0$. So a tangent vector sends any constant function to zero. Also if we have two functions f, g such that $f(p) = 0 = g(p)$, then for any tangent vector η to M at p we have $\eta(fg) = 0$ (by (ii)).

Any chart $\phi : U \to M$ (with $p \in \phi(U)$) in the differentiable structure gives some tangent vectors as follows. If f is in $A(p)$, then $f \circ \phi$ is a map of $U \subset R^n$ into R and we let

$$\partial_i(p)(f)$$

be the ith partial derivative of $f \circ \phi$ at $\phi^{-1}(p)$. Thus $\partial_i(p) : A(p) \to R$ is a tangent vector. So the chart (U, ϕ) gives n tangent vectors $\partial_1(p), \ldots, \partial_n(p)$ at p.

Let T_pM be the set of all tangent vectors to M at p. If we define operations by

$$(\xi+\eta)(f) = \xi(f) + \eta(f)$$

$$(r\xi)(f) = r\xi(f) \; .$$

then it is routine to verify that T_pM becomes a real vector space.

__Proposition__ 1: $\dim T_pM = n$ $(= \dim M)$.

<u>Proof</u>: We have seen that if we take a chart $\phi : U \to M$ in the differentiable structure with $p \in \phi(U)$, we get n tangent vectors $\partial_1(p), \ldots, \partial_n(p)$ in T_pM. We will show that these are a basis for T_pM.

We may assume $\phi(0) = p$. Then for any $f \in A(p)$ we can write

$$f(x) = f(p) + \sum_{i=1}^{n} \partial_i(p)(f)x_i + \sum_{i,j=1}^{n} \phi_{ij}(x)x_ix_j$$

for suitable smooth functions ϕ_{ij}. This is just the multivariate version of the following result about functions of one variable. Let f be a smooth real-valued function defined on some neighborhood of 0 in $\mathbb{R}$. We note that

$$\psi(x) = \frac{\frac{f(x)-f(0)}{x} - f'(0)}{x}$$

is smooth (and defined for $x \neq 0$). Solving for $f(x)$ gives

$$f(x) = f(0) + f'(0)x + \psi(x)x^2 .$$

Now let η be any element of T_pM and apply η to f. η is linear, so $\eta(f(p)) = 0$ since $f(p)$ is constant, and $\eta(\phi_{ij}(x)x_ix_j) = 0$ since both $\phi_{ij}(x)x_i$ and x_j vanish at p. Thus

$$\eta(f) = \eta(x_1)\partial_1(p)(f) + \ldots + \eta(x_n)\partial_n(p)(f) ,$$

and we conclude that

$$\eta = \eta(x_1)\partial_1(p) + \ldots + \eta(x_n)\partial_n(p) ,$$

showing that $\{\partial_1(p), \ldots, \partial_n(p)\}$ spans T_pM.

Now these tangent vectors are linearly independent, for if

$$c_1\partial_1(p) + \ldots + c_n\partial_n(p)$$

is the zero tangent vector and we apply it to the function x_i we get $c_i = 0$.

q.e.d.

Definition: A vector field X on an open set W in M is an assignment of $X_p \in T_pM$ for each $p \in W$.

We are interested in vector fields which are continuous in the sense that if points p and q are close then X_p and X_q are close. But since X_p and X_q are in different vector spaces, this looks difficult to formulate. Actually there is a neat way of even defining a smooth vector field.

Let X be a vector field on W and let $f : W \to \mathbb{R}$ be smooth. We get a new function

$$Xf : W \to \mathbb{R}$$

defined by $(Xf)(p) = X_p(f)$. This makes sense because $f \in A(p)$ for each $p \in W$.

Definition: The vector field X is smooth if for each smooth f, we have Xf smooth.

It is easy to verify that if X and Y are smooth vector fields then so is $X + Y$,

$$(X+Y)(f) = X(f) + Y(f) \quad .$$

Also if X is a smooth vector field and $r \in \mathbb{R}$, then rX is smooth $(rX(f) = r(X(f)))$. Thus the smooth vector fields form a vector space.

For a coordinate chart $\phi : U \to M$ in the differentiable structure we have smooth vector fields $\partial_1,\ldots,\partial_n$. So if $\sigma_1,\ldots,\sigma_n$ are smooth real-valued functions on $\phi(U)$ we have

$$\sigma_1\partial_1 +\ldots+ \sigma_n\partial_n$$

as a smooth vector field. Conversely, if X is smooth on U , we have

$$X = \rho_1\partial_1 +\ldots+ \rho_n\partial_n \quad ,$$

and the functions ρ_i are smooth because ρ_i is $X(x_i)$, the function obtained by applying the smooth vector field X to the smooth function x_i .

If we have two smooth vector fields X and Y we can get a real-valued operator on real-valued smooth functions as follows:

$$f \mapsto X_p(Yf) \quad .$$

Since Y and f are smooth, so is Yf and therefore X_p assigns the real number $X_p(Yf)$ to this function. But this operator from A(p) to $\mathbb{R}$ need not be a tangent vector. It is linear and depends only on f near p (condition (i)), but it may fail to satisfy condition (ii) in the definition of a tangent vector. Indeed, if $X = \partial_i$ and $Y = \partial_j$, then (ii) is false because the operation is just the mixed second partial derivative.

<u>Proposition</u> 2: <u>For smooth vector fields</u>, X,Y <u>the operator</u>

$$f \mapsto X_p(Yf) - Y_p(Xf)$$

<u>is a tangent vector</u>. ("The mixed second partials cancel out") .

<u>Proof</u>: Let X be a smooth field and f,g be smooth functions. We assert that

$$(+) \qquad X(fg) = (Xf)g + f(Xg) \quad .$$

At a point p the left hand side is

$$X_p(fg) = X_p(f)g(p) + f(p)X_p(g)$$

since X_p satisfies (ii), and this is just the right hand side evaluated at p. Of course, we have a similar formula for $Y(fg)$.

Then

$$X_p(Y(fg)) = X_p((Yf)g + f(Yg))$$

$$= X_p(Yf)g(p) + (Yf)(p)X_p(g) + X_p(f)(Yg)(p) + f(p)X_p(Yg) \quad ,$$

and similarly

$$Y_p(X(fg)) = Y_p(Xf)g(p) + (Xf)(p)Y_p(g) + Y_p(f)(Xg)(p) + f(p)Y_p(Xg) \quad .$$

Thus

$$(X_pY - Y_pX)(fg) = (X_pY - Y_pX)(f)g(p) + f(p)(X_pY - Y_pX)(g) \quad ,$$

proving that condition (ii) is satisfied. The linearity and condition (i) are true since they are true for both terms. So

$$X_pY - Y_pX \text{ is a tangent vector at } p \quad .$$

Let $[X,Y] = XY - YX$ be the vector field defined by $[X,Y]_p = X_pY - Y_pX$.

<u>Proposition</u> 3: <u>The set</u> $\mathfrak{X}(W)$ <u>of smooth vector fields on an open set</u> W <u>in a differentiable manifold</u> M <u>forms a Lie algebra under</u> $[\ ,\ ]$.

<u>Proof</u>: We have seen that $\mathfrak{X}(W)$ is a vector space. From the definition of $[X,Y]$ it is obvious that $[Y,X] = -[X,Y]$ and it is easy to show that this multiplication distributes over addition ($[X+Y,Z] = [X,Z] + [Y,Z]$) . Proof of the Jacobi identity is formally just the same as for matrix multiplication with $[A,B] = AB - BA$.

This Lie algebra $\mathfrak{X}(W)$ is usually infinite dimensional, but when our differentiable manifold is a Lie group we will get an important finite-dimensional subalgebra. (See section C).

Let M,N be differentiable manifolds and

$$M \xrightarrow{\psi} N$$

be a smooth map. If $\psi(p) = q$ we get a map

$$T_pM \xrightarrow{d\psi} T_qN$$

called the <u>differential</u> of ψ as follows:

If $\xi \in T_pM$ and $f \in A(q)$ we set

$$(\star) \qquad d\psi\xi(f) = \xi(f \circ \psi) .$$

<u>Proposition</u> 4: $(\star)$ <u>defines a map of</u> T_pM <u>into</u> T_qN <u>and it is a linear map</u>.

<u>Proof</u>: We need to show that $d\psi\xi$ is a linear derivation from $A(q)$ to $\mathbb{R}$ and assigns the same real number to two functions which agree on any neighborhood of q . This final condition is inherited

from the same property for ξ and linearity is easy to prove. So suppose $f,g \in A(q)$. We have

$$d\psi\xi(fg) = \xi(fg \circ \psi) = \xi((f \circ \psi)(g \circ \psi))$$

$$= (f \circ \psi)(p)\xi(g \circ \psi) + \xi(f \circ \psi)(g \circ \psi)(p)$$

$$= f(q)d\psi\xi(g) + d\psi\xi(f)g(q) \quad ,$$

showing that $d\psi\xi$ is a derivation.

Finally $d\psi : T_pM \to T_qN$ is linear. For we have

$$d\psi(a\xi+b\eta)(f) = (a\xi+b\eta)(f \circ \psi) = a\xi(f \circ \psi) + b\eta(f \circ \psi) = (ad\psi\xi+bd\psi\eta)(f) \; .$$

C. Lie groups

Let G be a differentiable n-manifold which is also a group and the operations

$$\begin{array}{ll} G \times G \to G & G \to G \\ (a,b) \mapsto ab & a \mapsto a^{-1} \end{array}$$

are smooth functions. Then G is called a Lie group.

Let G be a Lie group with identity element e, and suppose X_e is a tangent vector at e $(X_e \in T_eG)$. Then we can get a vector field defined on all of G as follows. For any $g \in G$ let $L_g : G \to G$ be the diffeomorphism given by $L_g(x) = gx$ for each $x \in G$. This is called left-translation by g. We set

$$(\dagger) \qquad X_g = dL_gX_e$$

$(dL_g : T_eG \to T_gG)$. Such vector fields are called <u>left-invariant</u>; i.e. a vector field X on G is left-invariant if it satisfies $(\dagger)$.

<u>Proposition</u> 5: <u>If</u> X,Y <u>are left-invariant vector fields on</u> G , <u>so is</u> $[X,Y]$.

<u>Proof</u>: Let $g \in G$ and $f \in A(g)$ and we calculate

$$dL_g[X,Y]_e(f) = [X,Y]_e(f \circ L_g)$$

$$= X_e(Y(f \circ L_g)) - Y_e(X(f \circ L_g))$$

$$= dL_gX_e(Yf) - dL_gY_e(Xf)$$

$$= X_g(Yf) - Y_g(Xf)$$

$$= [X,Y]_g(f) \text{ , proving } (\dagger) \text{ .}$$

Now one can see easily that X,Y left invariant implies $X+Y$ is also and so is rX $(r \in \mathbb{R})$. Thus the set of left-invariant fields on G becomes a subalgebra of the Lie algebra of all smooth vector fields. Since left-invariant vector fields correspond one-to-one with elements of T_eG , this Lie algebra is n-dimensional. We denote it by $\mathcal{L}(G)$ and call it the <u>Lie algebra of</u> G . It is convenient to use the language of categories and functors to discuss passing from G to $\mathcal{L}(G)$.

<u>Categories and Functors</u>

A category consists of <u>objects</u> $A,B,C,\ldots,$ and for each pair A,B of objects a set $\mathrm{Hom}(A,B)$ called the <u>morphisms</u> from A to B . Each $\mathrm{Hom}(A,A)$ is required to contain the identity morphism i_A , and there must be a <u>law of composition</u> so that $\alpha \in \text{Hom}(A,B)$ and $\beta \in \text{Hom}(B,C)$ give a unique $\beta \circ \alpha \in \text{Hom}(A,C)$.

<u>Examples</u>:

(i) objects - sets

morphisms - functions

composition is ordinary composition of functions

(ii) objects - groups

morphisms - homomorphisms

ordinary composition

(iii) objects - vector spaces

morphisms - linear maps

ordinary composition

(iv) objects - differentiable manifolds

morphisms - smooth maps

<u>Definition</u>: $\gamma \in \text{Hom}(A,B)$ is an <u>isomorphism</u> if there exists $\delta \in \text{Hom}(B,A)$ such that

$$\delta \circ \gamma = i_A \quad \text{and} \quad \gamma \circ \delta = i_B \ .$$

In (i) an isomorphism is a one-to-one function, in (ii) it is a group isomorphism, in (iii) it is a linear isomorphism and in (iv) it is a diffeomorphism.

If C_1 and C_2 are categories a <u>functor</u>

$$F : C_1 \to C_2$$

must send objects to objects and morphisms to morphisms such that:

we have either

$$\text{(I)} \qquad \begin{array}{ccc} A & \longrightarrow & F(A) \\ \alpha\downarrow & & \downarrow F(\alpha) \\ B & \longrightarrow & F(B) \\ \beta\downarrow & & \downarrow F(\beta) \\ C & \longrightarrow & F(C) \end{array} \qquad \text{with } F(\beta \circ \alpha) = F(\beta) \circ F(\alpha)$$

or

$$\text{(II)} \qquad \begin{array}{ccc} A & \longrightarrow & F(A) \\ \alpha\downarrow & & \uparrow F(\alpha) \\ B & \longrightarrow & F(B) \\ \beta\downarrow & & \uparrow F(\beta) \\ C & \longrightarrow & F(C) \end{array} \qquad \text{with } F(\beta \circ \alpha) = F(\alpha) \circ F(\beta)$$

In case (I) F is called a <u>covariant</u> functor and in case (II) it is a <u>contravariant</u> functor.

If C_1 and C_2 are both example (iii) above, then the functor which assigns to V its dual V^* gives a contravariant functor.

We have a category with Lie groups as objects and smooth homomorphisms as morphisms and we have a category of Lie algebras and Lie algebra homomorphisms.

<u>Proposition 6</u>: <u>The assignment</u> $G \to \mathfrak{L}(G)$ <u>gives a covariant functor from the Lie group category to the Lie algebra category</u>.

<u>Proof</u>: We must first say how a homomorphism of Lie groups induces a homomorphism of their Lie algebras. Let

$$G \xrightarrow{\phi} H$$

be a homomorphism of Lie groups. We get a linear map

$$T_eG \xrightarrow{d\phi} T_eH$$

by taking the differential of ϕ . Since these can be identified with $\mathfrak{L}(G)$ and $\mathfrak{L}(H)$ we have a linear map of the Lie algebras. We must just check that it is a homomorphism of Lie algebras; i.e. that it preserves the product $[X,Y]$.

Let $X,Y \in \mathfrak{L}(G)$ (i.e. left-invariant vector fields on G) and we calculate

$$d\phi[X,Y]f = [X,Y](f \circ \phi) = X(Y(f \circ \phi)) - Y(X(f \circ \phi))$$

$$= d\phi X(Yf) - d\phi Y(Xf) = [d\phi X, d\phi Y]f \quad .$$

Since this holds for all f , $d\phi[X,Y] = [d\phi X, d\phi Y]$ as required.

The functor $\mathfrak{L}$ does not map objects in a one-to-one manner -- two nonisomorphic Lie groups can have isomorphic Lie algebras. In fact, if $\phi : G \to H$ is a Lie group homomorphism which is a diffeomorphism on some neighborhood of e in G , the Lie algebra homomorphism $d\phi : \mathfrak{L}(G) \to \mathfrak{L}(H)$ will be an isomorphism of Lie algebras. Recall the two-to-one homomorphism ρ of $Sp(1)$ onto $SO(3)$. We see that $d\rho$ is an isomorphism of their Lie algebras (we proved these are isomorphic in Exercise 5 Chapter IV).

It turns out, however, that every real finite-dimensional Lie algebra is the Lie algebra of some Lie group. Indeed there is the following theorem (which we are not prepared to prove here - see page 133 of Hochschild, The Structure of Lie Groups, Holden-Day, 1965).

Theorem: *Let* $\mathcal{L}$ *be a finite dimensional real Lie algebra. Let* G *be the group of self-isomorphisms of* $\mathcal{L}$ *. Then* $\mathcal{L}(G) \cong \mathcal{L}$.

We conclude this section with some remarks about subgroups and subalgebras. We would like somehow to have a subgroup H of a Lie group G have its Lie algebra $\mathcal{L}(H)$ be a subalgebra of the Lie algebra $\mathcal{L}(G)$. The catch is that H may not be a Lie group at all. The simple classical example is: $G = S^1 \times S^1$ (a 2-torus). $G = \exp(R^2)$ where $\exp(x,y) = (e^{2\pi ix}, e^{2\pi iy})$. Let L be a line through $(0,0)$ in R^2 making angle θ with the x-axis where θ is chosen so that $\frac{2\pi}{\theta}$ is irrational. Let $H = \exp(L)$. This is a subgroup of R^2 , but is not a Lie group since it is not a manifold. It is *dense* in G (every point of G is a limit point of H). If, on the other hand, $\frac{2\pi}{\theta}$ is rational then H will be a circle subgroup of G . The result (which we will not prove here) is that a *closed* subgroup of a Lie group is itself a Lie group. For G and H as first described (H bad) we have that $\mathcal{L}(G) = R^2$ is the trivial Lie algebra (always $[X,Y] = 0$) and thus L is a Lie subalgebra of $\mathcal{L}(G)$. $\exp(L)$ is a subgroup H of G, but the subalgebra L cannot be $\mathcal{L}(H)$ because $\mathcal{L}(H)$ does not exist.

D. Connected groups

Proposition 7: *Let* X *be a pathwise connected space and let* Y *be a subset of* X *which is both open and closed. Then if* $Y \neq \phi$, $Y = X$.

<u>Proof</u>: We assume Y is not empty and choose $y \in Y$. For any $x \in X$ take a path $\rho : [0,1] \to X$ from y to x.

Let C be the points $t \in [0,1]$ which map into Y. We have $0 \in C$ (since $\rho(0) = y \in Y$), and we will show $1 \in C$ so that $x \in Y$.

Now C is closed. For any limit point of C maps to a limit point of Y (by continuity of ρ). This limit point of Y belongs to Y so, by definition of C, the limit point of C belongs to C.

If $1 \notin C$ let t_0 be the least upper bound of C. Since C is closed $t_0 \in C$, and thus $\rho(t_0) \in Y$. Since Y is open and ρ is continuous, some interval $(t_0 - \epsilon, t_0 + \epsilon)$ maps into Y, contradicting that t_0 is the least upper bound of C. Thus $1 \in C$ and $x \in Y$.

<u>Proposition</u> 8: <u>If G is a pathwise connected Lie group and H is a subgroup which contains an open neighborhood U of e in G, then</u> $H = G$.

<u>Proof</u>: Let

$$U^2 = \{xy \mid x \in U, y \in U\}$$

$$U^3 = \{xyz \mid x \in U, y \in U, z \in U\}, \text{ etc.}$$

Since H is a subgroup we see that

$$W = U \cup U^2 \cup U^3 \ldots$$

lies in H. As a union of open sets, W is an open set. But W is also closed, for suppose x is a limit point of W. Then the open set xU contains some

$$u_1u_2 \ldots u_m \in W \quad (\text{each } u_i \in U) \quad .$$

That is, there is $u \in U$ such that

$$xu = u_1u_2 \ldots u_m \, ,$$

but then $x = u_1u_2 \ldots u_m u^{-1} \in W$.

By Proposition 7 we see that W being both open and closed and nonempty implies that $W = G$. Since $W \subset H$ we have $H = G$.

<u>Corollary</u>: <u>Let</u> $f: K \to G$ <u>be a homomorphism of Lie groups with</u> G <u>pathwise connected</u>. <u>If</u> $f(K)$ <u>contains some open neighborhood of</u> e <u>in</u> G , <u>then</u> f <u>is surjective</u>.

If G is a Lie group which is not connected, let G_0 denote all $x \in G$ which can be connected to the identity element e by a path in G . We call G_0 the <u>identity component</u> of G .

<u>Proposition</u> 9: <u>The identity component</u> G_0 <u>of</u> G <u>is a pathwise connected subgroup of</u> G <u>which is both open and closed in</u> G .

<u>Proof</u>: G_0 is pathwise connected, for if $x,y \in G_0$ we can choose paths ρ,σ from e to x,y and then $\rho^{-1}\sigma$ is a path from x to y and it lies in G_0 since each $\rho(t)$ and $\sigma(t)$ belong to G_0 .

G_0 is a subgroup because it contains e and if $x,y \in G_0$ with paths ρ,σ then the path

$$(\rho\sigma)(t) = \rho(t)\sigma(t)$$

is a path in G from e to xy . Thus $xy \in G_0$. Similarly, if ρ is a path from e to x , then

$$\tau(t) = (\rho(t))^{-1}$$

is a path from e to x^{-1} .

G_0 is open. For if $x \in G_0$ we can take an open neighborhood U of x in G homeomorphic to an open ball in R^n (G is an n-manifold). Since x can be connected by a path ρ from e and each point in U can be connected by a path from x in U , we see that $U \subset G_0$.

G_0 is closed. For if x is a limit point of G_0 we can take U as above. Some point y in U is in G_0 and so we have a path σ from e to y . Since we have in U a path from y to x we see that $x \in G_0$.

We are going to need an <u>exponential</u> <u>map</u> from T_eG to G , and we cannot use the one from Chapter IV because the elements of T_eG and G cannot be assumed to be matrices. If $\gamma : (-\epsilon, \epsilon) \to G$ is a smooth curve with $\gamma(0) = e$ we define the <u>derivative</u> $\gamma'(0)$ of γ at e to be

$$\gamma'(0) = d\gamma(\xi)$$

where ξ is the unit vector 1 in $\mathbb{R}$ $(= T_0(\mathbb{R}))$.

We will need the following uniqueness theorem from differential-equation theory (see Milnor, Morse Theory, Annals of Math Studies #51, 1963, Lemma 2.4).

<u>Theorem</u>: <u>Given any</u> $\eta \in T_eG$, <u>there exists exactly one one-parameter subgroup</u> γ <u>in</u> G <u>such that</u>

$\gamma'(0) = \eta$. (see Chapter IV) .

<u>Definition</u>: Given any $\eta \in T_eG$, take γ to be the one-parameter subgroup in G such that $\gamma'(0) = \eta$ and set

$$\exp(\eta) = \gamma(1) \quad .$$

Let η and γ be as in this definition. Set

$$\rho(t) = \exp(t\eta) \quad .$$

We claim that this curve in G is just the one-parameter subgroup γ . To see this hold t fixed and consider the curve $\sigma(u) = \gamma(tu)$. Then σ is easily seen to be a one-parameter subgroup and

$$\sigma'(0) = t\gamma'(0) = t\eta \quad .$$

Thus, by definition of exp, $\exp(t\eta) = \sigma(1) = \gamma(t) = \rho(t)$. So we have proved

<u>Proposition</u> 10: *$t \mapsto \exp(t\eta)$ is the unique one-parameter subgroup whose derivative at 0 is η .*

This implies that exp when restricted to a one-dimensional subspace of T_eG is a homomorphism into G .

For matrix groups we saw that the exponential map was a diffeomorphism of some neighborhood of I in the group (having log as inverse). This is still true in our more general setting and is based on the

<u>Inverse Function Theorem</u>: *Let M,N be differentiable n-manifold and $\phi : M \to N$ be a smooth map. If $d\phi : T_pM \to T_{\phi(p)}N$ is an*

isomorphism, then ϕ *is a diffeomorphism of some open neighborhood of* p.

(A proof of this can be found in most good advanced calculus books).

We now apply this theorem to

$$\exp : T_eG \to G \quad .$$

We have $d(\exp) : T_eG \to T_eG$. Let $\eta \in T_eG$ and let γ be the unique one-parameter subgroup such that $\gamma'(0) = \eta$. Then (as we saw in the proof of Proposition 10) $\exp(t\eta) = \gamma(t)$ so that γ is the image under exp of the line $t \mapsto t\eta$. So $d(\exp)$ sends the tangent vector to this line (namely η) to $\gamma'(0)$ (which is also η) . Thus

Proposition 11: $d(\exp) : T_eG \to T_eG$ *is the identity map*.

Corollary: *If* G *is a pathwise connected Lie group, then* $\exp : T_eG \to T_eG$ *is surjective. If* G *is not pathwise connected,* $\exp T_eG$ *is the identity component* G_0 *of* G.

E. Abelian groups

Just as for abelian matrix groups we have the following result (and the proof is the same).

Proposition 12: *Let* G *be an abelian Lie group. If* γ, ρ *are one-parameter subgroups, then so is their product.*

Again, just as for abelian matrix groups we have

Proposition 13: Let G be an abelian Lie group. Then for any $X, Y \in T_eG$, we have

$$\exp(X+Y) = \exp(X)\exp(Y) \quad .$$

Proof: Let γ, ρ be the one-parameter subgroups satisfying $\gamma'(0) = X$, $\rho'(0) = Y$. By Proposition 12, $\gamma\rho$ is a one-parameter subgroup and $(\gamma\rho)'(0) = \gamma'(0)\rho(0) + \gamma(0)\rho'(0) = \gamma'(0) + \rho'(0) = X+Y$. Thus $\exp(X+Y) = (\gamma\rho)(1) = \gamma(1)\rho(1) = \exp(X)\exp(Y)$.

Thus for an abelian Lie group G, exp is a homomorphism from the additive vector group of T_eG to the group G. Taking note of the corollary to Proposition 11, we see that if G is pathwise connected, then exp is surjective. Thus a connected abelian group is a quotient group of the vector group T_eG.

Again, just as for matrix groups, exp is a diffeomorphism of some neighborhood of 0 in T_eG so that the kernel is a discrete subgroup of the vector group T_eG. So just as for matrix groups

Theorem: A connected abelian Lie group G is isomorphic to a cartesian product of a torus and a vector group. If G is also compact, it is a torus.

Chapter 13
Reflections, Roots

A. Maximal tori

For our matix groups we proved that if G is compact and connected and T is a maximal torus in G, then

(i) $G = \bigcup_{x \in G} xTx^{-1}$, and

(ii) Any maximal torus in G is some xTx^{-1}.

The proof we used for (ii) (pp. 126, 127) applies directly in our more general setting. But the proofs we used for (i) depended strongly on properties of matrices - indeed we gave different proofs for $SU(n)$, $SO(n)$, and $Sp(n)$. We will prove (i) later, but for now we assume it.

Definition: A group G is divisible if given $x \in G$ and any positive integer m, there exists $y \in G$ such that $y^m = x$.

Example: The group S^1 of complex numbers of unit length is divisible. If G and H are divisible, then so is $G \times H$. Thus any torus is a divisible group.

<u>Proposition</u> 1: <u>Let</u> G <u>be an abelian Lie group whose identity component</u> T <u>is a torus</u>. <u>If</u> G/T <u>is finite cyclic</u>, <u>then</u> G <u>is monogenic</u>.

<u>Proof</u>: We proved (page 126) that T is monogenic, and we begin by choosing a generator x for T. Next we choose $y \in G$ such that its image in G/T generates G/T, which is isomorphic, say, with $\mathbb{Z}m$. Then $y^m \in T$, and thus so are y^{-m} and xy^{-m}

Now using the fact that T is divisible, we can choose $z \in T$ such that

$$z^m = xy^{-m}.$$

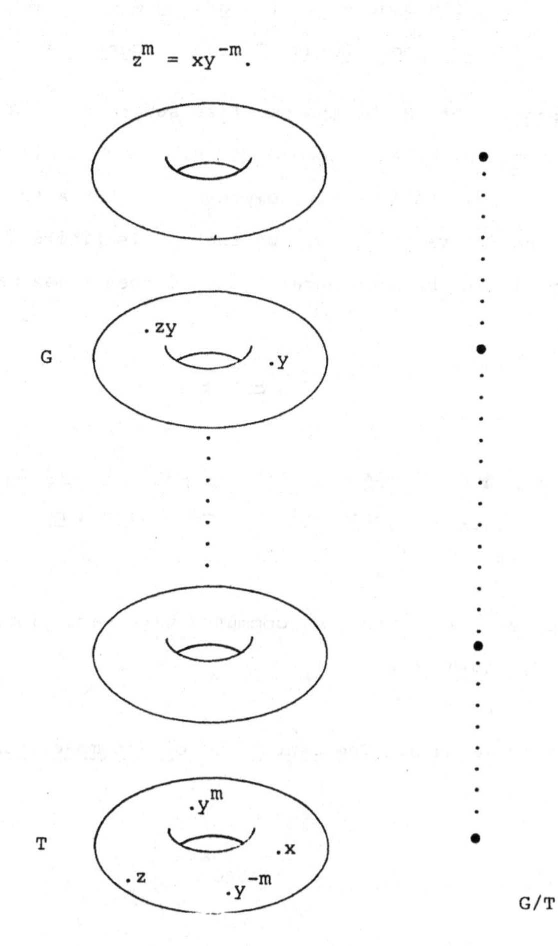

Since G is abelian, this is the same as

$$(zy)^m = z^m y^m = x.$$

We claim that zy generates G. Since $(zy)^m = x$ and x generates T, the powers of zy are dense in T. Since $z \in T$, for $r = 1, \ldots, m$ we have that $(zy)^r = z^r y^r$ is in Ty^r, so the powers of zy are also dense in the coset Ty^r; thus in all of G.

__Proposition__ 2: __Let__ G __be a compact connected Lie group,__ S __a connected abelian subgroup. If some__ $g \in G$ __commutes with each element of__ S, __then some torus__ T __in__ G __contains__ $S \cup g$.

__Proof__: Let H be the smallest subgroup of G containing $S \cup g$. Let $\bar{H}$ be the closure of H. Then $\bar{H}$ is compact and abelian. Thus its identity component $\bar{H}^\circ$ is a torus. $\bar{H}/\bar{H}^\circ$ is finite and generated by g, so that it is finite cyclic. By Proposition 1, $\bar{H}$ has a generator h. Choose a maximal torus T containing h. Then

$$g \cup S \subset H \subset \bar{H} \subset T.$$

__Corollary__: __"A maximal torus is maximal abelian" Precisely, if__ A __is an abelian subgroup of__ G __containing a maximal torus__ T, __then__ $A = T$

We see that if $g \in G$ commutes with each element of a maximal torus T, then $g \in T$.

Proposition 3: __The center__ C __of a compact connected Lie group__ G __is given by__

$$C = \bigcap_{x \in G} xTx^{-1}$$

Proof: By the observations above any $c \in C$ must lie in each maximal torus. Conversely, if $y \in G$ belongs to each maximal torus, then, since these cover G, y commutes with every element of G.

q.e.d.

Given $g \in G$, what can we say about its centralizer

$$C(g) = \{x \in G \mid xg = gx\} \ ?$$

Well, $C(g)$ may fail to be connected, but its identity component $C^\circ(g)$ is easy to describe.

Proposition 4: $C^\circ(g)$ is the union of all maximal tori which contain g.

Proof: If T is a maximal torus containing g then clearly $T \subset C^\circ(g)$.

Conversely, suppose $x \in C^\circ(g)$. The x belongs to some maximal torus S of $C^\circ(g)$. And some maximal torus T in G contains $S \cup g$.

Definition: An element x of a compact connected Lie group G is regular if it belongs to exactly one maximal torus. Otherwise it is singular.

If G is a torus, then all elements are regular. Otherwise, the identity element is singular.

Example: Let T be a maximal torus, N its normalizer and $W = N/T$ be the Weyl group.

$$0 \to T \overset{\alpha}{\to} N \overset{\beta}{\to} W \to 1$$

We know that x, y in N have the same action on T (by conjugation) $\Longleftrightarrow \beta(x) = \beta(y)$.

Now suppose $w \in W$ is of order 2; i.e. $w \neq 1$, but $w^2 = 1$. Then for any $x \in \beta^{-1}(w)$ we easily see that $x^2 \in T$ is a singular element, because $w \neq 1$, and we see that $x \notin T$. Thus there is some

$$yTy^{-1} \neq T$$

with $x \in yTy^{-1}$. But then $x^2 \in yTy^{-1} \cap T$ and thus is singular.

We now get our first indication that a lot of information about G is contained in N.

Proposition 5: *If $t, t' \in T$ are conjugate in G, then they are conjugate in N.*

Proof: We are given that

$$gtg^{-1} = t'$$

for some $g \in G$ and we must find $x \in N$ such that $xtx^{-1} = t'$.

Let C, C' be the identity components of the centralizers of t, t'. These are characterized in Proposition 4.

Now $t' \in gTg^{-1} \subset C'$ and $T \subset C'$ so that gTg^{-1} and T are two maximal tori in the Lie group C'. Thus there exists $h \in C'$ such that

$$h(gTg^{-1})h^{-1} = T.$$

Then $x = hg$ normalizes T; i.e. $x \in N$. Also

$$xtx^{-1} = hgtg^{-1}h^{-1} = ht'h^{-1} = t'.$$

q.e.d.

Corollary: *If $z \in G$ commutes with each $x \in N$, then z commutes with each $y \in G$ (i.e. $z \in C(G)$).*

For each $w \in W$ we have the action $t \to wt$ (lift to N & conjugate) and we let $V(w)$ be the fixed-point set of this action

$$V(w) = \{t \ T | wt=t\}.$$

<u>Proposition</u> 6: Center $G = \bigcap_{w \in W} V(w)$.

<u>Proof</u>: If $z \in$ Center G then conjugation of z by any element of G leaves it fixed, so, a fortiori, this is true of each element of N. Conversely, suppose z belongs to each $V(w)$. Then z commutes with every element of N. By Proposition 5, z commutes with every element of G.

B. <u>The anatomy of a reflection</u>

Here, as usual, G is a compact connected Lie group, T is a maximal torus in G, N is the normalizer of T in G and $W = N/T$ is the Weyl group.

$$0 \to T \to N \overset{\rho}{\to} W \to 1.$$

Also as usual, W acts on T by: for $w \in W$ choose $x \in \rho^{-1}(w) \subset N$ and conjugate T by x. Since T is abelian this is independent of choice. The action of each $w \in W$ is a group isomorphism which is also a diffeomorphism. Thus the differential dw of w is a linear isomorphism of the tangent space $L(T)$ to T at the identity.

<u>Definition</u>: $w \in W$ is a <u>reflection</u> if $dw : L(T) \to L(T)$ is a reflection in an $(r-1)$-dimensional subspace of the r-dimensional real vector space $L(T)$.

Given a reflection $w \in W$ we let <u>$K(w)$</u> be the <u>fixed-point set</u> of w; i.e.

$$K(w) = \{t \in T | w(t)=t\}.$$

Clearly $K(w)$ is a closed subgroup of T and, since w is a reflection, dim $K(w) = r-1$. Similarly, we set

$$L(w) = \{t \in T \mid w(t) = t^{-1}\}$$

and note that $L(w)$ is also a closed subgroup of T. Clearly every element of $K(w) \cap L(w)$ is a square root of unity in T and thus $K(w) \cap L(w)$ is a finite set. Since dim $K(w) = r-1$, it follows that dim $L(w) = 1$. Thus $L(w)$ is topologically a finite disjoint union of circles.

Definition: Let $Q(w)$ be the identity component of the group $L(w)$. This circle subgroup $Q(w)$ of T is called the coroot associated with the reflection w.

Now $Q(w)$ contains the identity element 1 (of G and of T) and $Q(w)$ also contains exactly one other square root of 1, and we denote this element by α (more properly, we should write $\alpha(w)$).

Proposition 7: $K(w)$ has at most two components.

Proof: If the $(r-1)$-dimensional subgroup $K(w)$ of T had more than two components (which are $(r-1)$-tori in T). Then $Q(w) \cap K(w)$ would have more than two points. But each point of $Q(w) \cap K(w)$ is a square root of unity in the circle group $Q(w)$, and there are only two of these.

q.e.d.

Now let w be a fixed reflection. Then we write K, Q, α, etc. without mentioning w. Let $\phi : G \to G$ be the squaring map $\phi(x) = x^2$. Let $T' = \rho^{-1}(w)$.

Proposition 8: $\phi(T')$ is a component of K.

Proof: For $x \in T'$ we have $x\phi(x)x^{-1} = x x^2 x^{-1} = x^2 = \phi(x)$, which shows that $\phi(T') \subset K$.

Next, we look at the "fiber" ϕ^{-1}(point)of ϕ. Suppose $x,y \in T'$ and $\phi(x) = \phi(y)$; i.e. $x^2 = y^2$. Then

$$x(xy^{-1})x^{-1} = x^2y^{-1}x^{-1} = y^2y^{1-}x^{-1} = yx^{-1} = (xy^{-1})^{-1},$$

which shows $xy^{-1} \in Q$. These steps reverse, and we see that the fibers of ϕ are cosets of Q. Thus $\phi(T')$ is a closed connected $(r-1)$-dimensional submanifold of K. Thus $\phi(T')$ is an $(r-1)$-torus in K and so is a component of K.

q.e.d.

<u>Proposition</u> 9: The identity component K° of K consists entirely of singular elements.

<u>Proof</u>: Now any $x \in T'$ commutes with all elements of K (by definition of K). In particular, $x \in T'$ commutes with all elements of the torus K° and by Proposition 2, there is some maximal torus zTz^{-1} containing both x and K°. Then

$$K^\circ \subset T \cap zTz^{-1}$$

(and $zTz^{-1} \neq T$ since $x \notin T$ and $x \in zTz^{-1}$).

<u>Definition</u>: We set

$$U = K^\circ \cup \phi(T')$$

and call U the <u>root</u> <u>kernel</u> associated with the reflection w. (Sometimes $\phi(T') = K^\circ$).

Now the root kernel U is either an $(r-1)$-torus or a $\mathbb{Z}_2$ extension of an $(r-1)$-torus. By Proposition 1, U is monogenic. We choose a generator u of U, and let P be the <u>identity</u> <u>component</u> <u>of</u> <u>the</u> <u>centralizer</u> of u. Set $H = \bigcup_{x \in P} xQx^{-1}$.

<u>Proposition</u> 10: H is a subgroup of P with the circle group Q as its maximal torus.

<u>Proof</u>: Clearly H is a subset of P but it is not obvious that it is a subgroup. We easily see that it contains the identity and if $y \in H$ then $y^{-1} \in H$. We need to see that H is closed under multiplication.

Consider the natural homomorphism

$$\eta : P \to P/U.$$

If the square root α in Q is not in U, then η is $1-1$ on each xQx^{-1} in H. If $\alpha \in U$, then η is $2-1$ on each xQx^{-1}. In either case $\eta(Q)$ is a maximal torus in the compact connected Lie group P/U. So conjugates of $\eta(Q)$ cover P/U.

Suppose xqx^{-1} and $yq'y^{-1}$ are elements of H. Then

$$\eta(xqx^{-1}\, yq'y^{-1}) = \eta(x)\,\eta(q)\,\eta(x^{-1})\,\eta(y)\,\eta(q')\,\eta(y^{-1})$$

lies in some $z\,\eta(Q)z^{-1}$ so that

$$xqx^{-1}\, yq'y^{-1} \in zQz^{-1} \subset H.$$

Clearly, Q is a torus in H. Since $Q \subset T$ we have $Q \subset H \cap T$.

C. <u>The adjoint representation</u>

Recall that we have an exponential map

$$\exp : L(G) \to G$$

defined as follows. For $\eta \in L(G)$ we let γ be the unique one-parameter subgroup of G such that $\gamma'(0) = \eta$. Then $\exp(\eta) = \gamma(1)$.

<u>Proposition 12: Let $\phi : G \to H$ be a homomorphism of Lie groups and $d\phi : L(G) \to L(H)$ be the corresponding Lie algebra homomorphism. Then the following diagram commutes.</u>

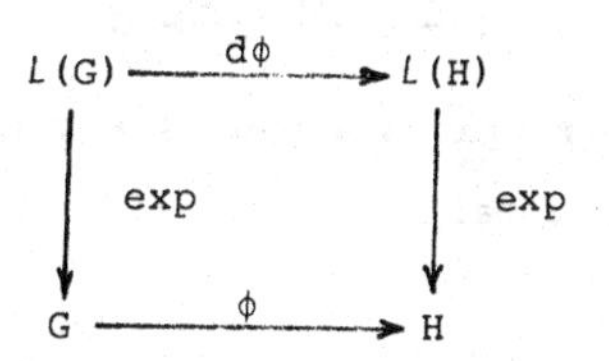

<u>Proof</u>: Let η and γ be as above. By the definition of $d\phi$, $d\phi(\eta)$ is the derivative at 0 of the one-parameter subgroup $\phi\circ\gamma$ of H; i.e. $d\phi(\eta) = (\phi\circ\gamma)'(0)$. Then

$$\phi(\exp\eta) = \phi(\gamma(1)) = (\phi\circ\gamma)(1) = \exp d\phi(\eta).$$

q.e.d.

For $x \in G$ we denote by

$$Ax : G \to G$$

the isomorphism of G onto itself given by

$$Ax(y) = xyx^{-1}.$$

Then its differential

$$dAx : L(G) \to L(G)$$

is a Lie algebra isomorphism. In particular it is an element of the general linear group $GL(L(G))$ and this group is just $GL(n, \mathbb{R})$ where $n = \dim G$.

By Proposition 12 we have a commutative diagram

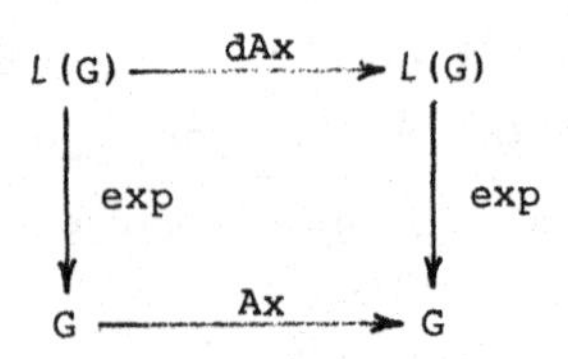

The assignment $x \to dAx$ gives a map $G \to GL(n, \mathbb{R})$. We agree to write $Ad(x)$ for dAx; i.e. we call this map (of G into $GL(n, \mathbb{R})$) Ad.

Proposition 13: *The map* $Ad : G \to GL(n, \mathbb{R})$ *is a group homomorphism and its kernel is the center of* G.

Proof:

$$Ad(xy) = d\,Axy = d(Ax \circ Ay) = d\,Ax \circ d\,Ay = Ad(x)\,Ad(y).$$

For the second assertion we note that $Ad(x) = dAx$ is the identity $\iff$ Ax is the identity $\iff$ x is in the center of G.

Now we have another commutative diagram

$$\begin{array}{ccc} (G) & \xrightarrow{d(Ad)} & Mn(\mathbb{R}) = (GL(n, \mathbb{R})) \\ \downarrow \text{exp} & & \downarrow e \\ G & \xrightarrow{Ad} & GL(n, \mathbb{R}) \end{array}$$

where e is exponentiation of matrices.

Notation: We write ad for $d(Ad)$

Definition: By the *adjoint representation* we mean Ad restricted to a maximal torus T in G

$$Ad : T \to GL(n, \mathbb{R}).$$

It is through this representation that we are going to define *the roots of* G. These roots are homomophisms of T to S^1 and the plan is to show that $Ad(T)$ lies in some "standard" torus T' in $GL(n, \mathbb{R})$ where T' has obvious projections $\pi_i : T' = S^1 \times \cdots \times S^1$. Then the roots will be $\theta_i = \pi_i \circ Ad$.

Recall that the standard maximal torus T' for $SO(n)$ is:

$$\begin{pmatrix} \text{rot } \theta_1 & & & 0 \\ & \text{rot } \theta_2 & & \\ & & \ddots & \\ 0 & & & \text{rot } \theta_k \end{pmatrix} \quad \text{or} \quad \begin{pmatrix} \text{rot } \theta_1 & & & & 0 \\ & \text{rot } \theta_2 & & & \\ & & \ddots & & \\ & & & \text{rot } \theta_k & \\ 0 & & & & 1 \end{pmatrix}$$

for $n = 2k$ for $n = 2k+1$

Here $\text{rot}\,\theta_i$ stands for the 2×2 block

$$\begin{pmatrix} \cos \theta_i & \sin \theta_i \\ -\sin \theta_i & \cos \theta_i \end{pmatrix}.$$

The projection π_i sends $t \in T'$ to $\theta_i(t)$.

Thus to get roots it suffices to show that $Ad(T) \subset T'$ where T' is the standard maximal torus in $SO(n)$. Now Ad is a homomorphism and is continuous so $Ad(T)$ is compact connected and is an abelian group. So $Ad(T)$ is a torus. If $Ad(T)$ lies in $SO(n)$ then $Ad(T)$ lies in a maximal torus in $SO(n)$. So (conjugating the representation if need be) we have $Ad(T) \subset T'$ and the roots are defined.

This is easily done using a little representation theory (see [1] page 82 for example), but we give a more computational "matrix group" argument instead.

If $A \in GL(n, \mathbb{R})$ commutes with all elements t of the maximal torus T' in $SO(n)$, then it is easy to prove that A is block diagonal with 2×2 blocks. (e.g. if $n = 4$ and $a_{13} \neq 0$ take

$$t = \begin{pmatrix} 0 & 0 & 0 & 0 \\ 0 & 0 & 0 & 0 \\ 0 & 0 & 1 & 0 \\ 0 & 0 & 0 & 1 \end{pmatrix}$$

and note that tA has 0 in the 1,3 position whereas At has $a_{13} \neq 0$ in the 1,3 position.

So we are reduced to the 2×2 case. Let

$$A = \begin{pmatrix} a & b \\ c & d \end{pmatrix} \quad \text{with} \quad ad - bc = r \neq 0.$$

Setting

$$\begin{pmatrix} a & b \\ c & d \end{pmatrix} \begin{pmatrix} \cos\theta & \sin\theta \\ -\sin\theta & \cos\theta \end{pmatrix} = \begin{pmatrix} \cos\theta & \sin\theta \\ -\sin\theta & \cos\theta \end{pmatrix} \begin{pmatrix} a & b \\ c & d \end{pmatrix}$$

easily gives $A \in SO(2)$.

So we have $Ad(T) \subset T'$ so for $t \in T$

$$Ad(t) = \begin{pmatrix} \text{rot } \theta_1(t) & & & \\ & \boxed{\text{rot } \theta_2(t)} & & \\ & & \ddots & \\ & & & \boxed{\text{rot } \theta_2(t)} \\ & & & & 1 \end{pmatrix}$$

(1 if n is odd)

The homomorphisms

$$\theta_i : T \to S^1$$

are the <u>roots</u> of G.

Let Vo be the subspace of $L(G)$ consisting of all vectors in $L(G)$ which are left fixed by all $Ad(t)$ $(t \in T)$.

Proposition 14: $Vo = L(T)$.

Proof: Clearly $L(T) \subset Vo$ (since T acts trivially on itself by conjugation).

Suppose $\eta \in Vo - L(T)$. Let γ be the one-parameter subgroup given by

$$\gamma(u) = \exp(u\eta).$$

For each $t \in T$ we see that conjugation by t leaves each $\gamma(u)$ fixed. For we have

$$\begin{aligned} A_t\, \gamma(u) &= A_t(\exp u\eta) = \exp \mathrm{Ad}(t)(u\eta) \\ &= \exp u\, \mathrm{Ad}(t)(\eta) = \exp u\eta \quad (\text{since } \eta \in Vo) \\ &= \gamma(u). \end{aligned}$$

Thus the subgroup H of G generated by T and $\{\gamma(u) \mid u \in \mathbb{R}\}$ is abelian. It is also connected and contains T. Since T is maximal, we must have $H = T$. But if η is a nonzero vector not in $L(T)$ we could not have all $\gamma(u) \in T$. Thus the proposition is proved.

Corollary: Let G have dimension n and rank r. Then

$$L(G) = L(T) \oplus V_1 \oplus \cdots \oplus V_s$$

where $V_1, \cdots, V_s$ are certain 2-planes on each of which some $\theta_i(t)$ is not the identity.

Proof: First off because of the block diagonal form for each $\mathrm{Ad}(t)$ we have

$$L(G) = P_1 \oplus P_2 \oplus \cdots \oplus P_k \quad (\oplus \mathbb{R} \text{ if } n \text{ is odd})$$

(here $k = [\frac{n}{2}]$ i.e. the biggest integer in $\frac{n}{2}$.) $\mathrm{Ad}(t)$ rotates P_1 by an angle $\theta_1(t)$, etc. Now if n is odd $\mathbb{R}$ is in $Vo = L(T)$, the subspace in which each $Ad(t)$ is the identity. So the 2-planes left over are the $V_1,\cdots,V_s$. Note that $n-r$ is always even.

D. Sample computation of roots

We compute the roots of some low-dimensional matrix groups.

(1) $G = SO(3)$

$$L(G) = so(3) = \{3\times 3 \text{ real skew-symmetric matrices}\}$$
$$\text{with } [A,B] = AB - BA$$

The standard maximal torus is

$$T = \left\{\begin{pmatrix} \cos\theta & \sin\theta & 0 \\ -\sin\theta & \cos\theta & 0 \\ 0 & 0 & 1 \end{pmatrix}\right\}.$$

As a basis for $L(G)$ we take

$$e_1 = \begin{pmatrix} 0 & 1 & 0 \\ -1 & 0 & 0 \\ 0 & 0 & 0 \end{pmatrix} \quad e_2 = \begin{pmatrix} 0 & 0 & 1 \\ 0 & 0 & 0 \\ -1 & 0 & 0 \end{pmatrix}$$

$$e_3 = \begin{pmatrix} 0 & 0 & 0 \\ 0 & 0 & 1 \\ 0 & -1 & 0 \end{pmatrix}.$$

Clearly $L(T) = \mathbb{R}e_1$ and we have

$$L(G) = L(T) \oplus V$$

where V is the span of e_2 and e_3.

Next we take one-parameter subgroups

$$\gamma(u) = \begin{pmatrix} \cos u & 0 & \sin u \\ 0 & 1 & 0 \\ -\sin u & 0 & \cos u \end{pmatrix}$$

$$\sigma(u) = \begin{pmatrix} 1 & 0 & 0 \\ 0 & \cos u & \sin u \\ 0 & -\sin u & \cos u \end{pmatrix}.$$

Then $\gamma'(0) = e_2$ and $\sigma'(0) = e_3$.

Now for

$$t = \begin{pmatrix} \cos\theta & \sin\theta & 0 \\ -\sin\theta & \cos\theta & 0 \\ 0 & 0 & 1 \end{pmatrix} \in T$$

we can calculate the action of $Ad(t)$ on V.

Let $\Gamma(u) = t\gamma(u)t^{-1}$, so that $Ad(t)e_2 = \Gamma'(0)$. This comes out to be

$$\begin{pmatrix} 0 & 0 & \cos\theta \\ 0 & 0 & -\sin\theta \\ -\cos\theta & \sin\theta & 0 \end{pmatrix} = (\cos\theta)\, e_2 - (\sin\theta)\, e_3.$$

Also, letting $\Sigma(u) = t\,\sigma(u)t^{-1}$ we get

$$Ad(t)\; e_3 = (\sin\theta)\, e_2 + (\cos\theta)\, e_3.$$

Thus $Ad(t)$ rotates V by an angle $-\theta$. We conclude that $SO(3)$ has one root, and we denote it by $-\theta$.

(2) $G = Sp(1)$

The Lie algebra $L(G)$ is the span of i,j,k in $\mathbb{H} = \mathbb{R}^4$. The standard maximal torus is

$$T = \{a + ib \mid a^2 + b^2 = 1\} = \{\cos\theta + i\sin\theta\}.$$

Then $L(T) = \mathbb{R}i$ (and, of course, $\mathrm{Ad}(t)i = i$).

Just as for $SO(3)$ we have

$$L(G) = L(T) \oplus V,$$

and for $Sp(1)$, V is the span of j and k.

$$\text{Let } \gamma(u) = \cos u + j \sin u \quad \text{and}$$
$$\sigma(u) = \cos u + k \sin u .$$

Then $\gamma'(0) = j$ and $\sigma'(0) = k$ and we calculate that

$$\begin{aligned}\mathrm{Ad}(t)j &= (t\gamma(u)t^{-1})(0) \\ &= j(\cos^2\theta - \sin^2\theta) + k(2\cos\theta \ \sin\theta) \\ &= j\cos 2\theta + k\sin 2\theta .\end{aligned}$$

Thus, for $t = \cos\theta + i\sin\theta$, we have that $\mathrm{Ad}(t)$ rotates the plane V by 2θ. So $Sp(1)$ has one root, 2θ.

(3) $G = U(2)$

G has dimension 4 and

$$L(G) = \{A \in M_2(\mathbb{C}) \mid A + {}^t\bar{A} = 0\}.$$

The rank of G is 2 and a standard maximal torus is

$$T = \left\{\begin{pmatrix} e^{i\theta_1} & 0 \\ 0 & e^{i\theta_2}\end{pmatrix}\right\}$$

As a basis for $L(G)$ we take:

$$e_1 = \begin{pmatrix} i & 0 \\ 0 & 0\end{pmatrix} \quad e_2 = \begin{pmatrix} 0 & 0 \\ 0 & i\end{pmatrix} \quad e_3 = \begin{pmatrix} 0 & 1 \\ -1 & 0\end{pmatrix} \quad e_4 = \begin{pmatrix} 0 & i \\ i & 0\end{pmatrix}$$

Note that e_1, e_2 are in $L(T)$ so that any $\mathrm{Ad}(t)$ is the identity on $L(T) = \mathrm{Span}(e_1, e_2)$.

So we calculate $Ad(t)$ on $V = \text{Span}(e_3, e_4)$.

$$\text{Let } \gamma(u) = \begin{pmatrix} 0 & \sin u + i\cos u \\ -\sin u + i\cos u & 0 \end{pmatrix} \text{ so that}$$

$\gamma'(0) = e_3$. Then a routine calculation gives

$$Ad(t)e_3 = (t\gamma(u)t^{-1})'(0) = \begin{pmatrix} 0 & e^{i(\theta_1-\theta_2)} \\ -e^{-i(\theta_1-\theta_2)} & 0 \end{pmatrix}$$

$$= \cos(\theta_1 - \theta_2)e_3 + \sin(\theta_1 - \theta_2)e_4.$$

Thus $U(2)$ has <u>one root</u>, $\theta_1 - \theta_2$.

(We have $\dim T = 2$, but have only one root. Basically this is because the center of $U(2)$ is a circle group and the center always goes to zero under Ad.)

(4) $G = U(3)$

Just as for $U(2)$, we have a circle in the center and do not expect the roots to span $L(T)^*$. The standard maximal torus for $U(3)$ is

$$T = \left\{ \begin{pmatrix} e^{i\theta_1} & 0 & 0 \\ 0 & e^{i\theta_2} & 0 \\ 0 & 0 & e^{i\theta_3} \end{pmatrix} \right\}$$

We take

$$e_1 = \begin{pmatrix} i & 0 & 0 \\ 0 & 0 & 0 \\ 0 & 0 & 0 \end{pmatrix} \quad e_2 = \begin{pmatrix} 0 & 0 & 0 \\ 0 & i & 0 \\ 0 & 0 & 0 \end{pmatrix} \quad e_3 = \begin{pmatrix} 0 & 0 & 0 \\ 0 & 0 & 0 \\ 0 & 0 & i \end{pmatrix}$$

as a basis for $L(T)$. We need six more basis vectors for $L(G)$ and we take

$$e_4 = \begin{pmatrix} 0 & 1 & 0 \\ -1 & 0 & 0 \\ 0 & 0 & 0 \end{pmatrix} \quad e_5 = \begin{pmatrix} 0 & 0 & 1 \\ 0 & 0 & 0 \\ -1 & 0 & 0 \end{pmatrix} \quad e_6 = \begin{pmatrix} 0 & 0 & 0 \\ 0 & 0 & 1 \\ 0 & -1 & 0 \end{pmatrix}$$

$$e_7 = \begin{pmatrix} 0 & i & 0 \\ i & 0 & 0 \\ 0 & 0 & 0 \end{pmatrix} \quad e_8 = \begin{pmatrix} 0 & 0 & i \\ 0 & 0 & 0 \\ i & 0 & 0 \end{pmatrix} \quad e_9 = \begin{pmatrix} 0 & 0 & 0 \\ 0 & 0 & i \\ 0 & i & 0 \end{pmatrix}.$$

The same kind of calculation as before gives us three (linearly dependent) roots

$$\theta_1 - \theta_2,\ \theta_2 - \theta_3,\ \theta_1 - \theta_3.$$

Appendix 1

In Exercise 8 of Chapter VI the concept of <u>limit point</u> is defined and it is proved that C is closed $\Leftrightarrow$ (x lp $C \Rightarrow x \in C$) ; i.e., a set is closed if and only if it contains all of its limit points.

<u>Lemma</u>: <u>If</u> $C \subset R^n$ <u>is bounded and infinite then</u> C <u>has a limit point</u>.

<u>Proof</u>: We must find $x \in R^n$ such that each $B(x,\varepsilon) \cap C$ is an infinite set.

Since C is bounded we can find an n - dimensional cube K_1 containing C. Let λ be the length of the sides of K_1. Divide each edge into two equal parts and cut K_1 into 2^n equal cubes of side length $\frac{\lambda}{2}$. Call those cubes $K_{11}, K_{12}, \ldots, K_{12^n}$. At least one of those must contain infinitely-many points of C. Choose one such and call it K_2. Subdivide K_2 into 2^n cubes of side $\frac{\lambda}{4}$ and choose one of these, K_3, containing infinitely-many points of C. Continue this process to get $K_1 \supset K_2 \supset K_3 \supset \ldots$. Clearly $K_1 \cap K_2 \cap K_3 \cap \ldots$ is a single point $x \in R^n$.

Then x lp C. For, consider any $B(x,\varepsilon)$. Take m such that $\frac{\lambda}{2^m} < \frac{\varepsilon}{2}$. Then

$$K_m \subset B(x,\varepsilon)$$

and K_m contains infinitely-many points of C. q.e.d.

<u>Theorem</u>: (Proposition 5 of Chapter VI) <u>If</u> $C \subset R^n$ <u>is closed and bounded and</u>

$$f: C \to R^m$$

<u>is continuous, then</u> $f(C)$ <u>is closed and bounded.</u>

<u>Proof</u>: Suppose $f(C)$ is not bounded. Choose $x_1 \in C$ such that $y_1 = f(x_1) \notin B(o,1)$.

Choose $x_2 \in C$ such that $y_2 = f(x_2) \notin B(o,2)$.

⋮

Choose $x_k \in C$ such that $y_k = f(x_k) \notin B(o,k)$. It is easy to prove that $Y = \{y_1,y_2,...\}$ has no limit point and that $X = \{x_1,x_2,...\}$ is an ifinite set. Also $X \subset C$ is bounded. So by the lemma X has some limit point x . Then also x lp C . Since C is closed $x \in C$. But then $y = f(x)$ must be a limit point of $Y = f(X)$ by continuity. Thus $f(C)$ is bounded.

Suppose $f(C)$ is not closed. Then there exists $y \in R^m$ such that y lp $f(C)$ but $y \notin f(C)$. Choose $x_1 \in C$ such that $y_1 = f(x_1) \in B(y,1)$.

Choose $x_2 \in C$ such that $y_2 = f(x_2) \in B(y,\frac{1}{2})$.

⋮

Choose $x_k \in C$ such that $y_k = f(x_k) \in B(y,\frac{1}{2})$. Clearly y is the only limit point of $Y = \{y_1,...,y_k,...\}$. Just as before $X = \{x_1,...,x_k,...\}$ is infinite and bounded. Let x lp X . Then x lp C so $x \in C$ and $f(x)$ lp $Y = f(X)$. This implies $f(x) = y$, but $f(x) \in f(C)$. So $f(C)$ is closed. q.e.d.

Appendix 2

Proof of Proposition 10 in Chapter I.

Let A be a finite-dimensional algebra (over some field k) and let $a \in A$. Then

$$\{a^0 = 1, a, a^2, a^3, \ldots\}$$

cannot all be linearly independent, so there is some polynomial $g(x) \in k[x]$ such that $g(x) = 0$. The monic polynomial with minimal degree having a as a zero is the minimal polynomial for a.

Lemma: If $a \in A$ and $p(x)$ is the minimal polynomial for a, then a is a unit in A <=> $p(x)$ has non-zero constant term.

Proof:

Let $p(x) = x^n + c_{n-1}x^{n-1} + \ldots c_1x + c_0$

<== If $c_0 \neq 0$ we can easily calculate that

$$(-1/c_0)(a^{n-1} + c_{n-1}a^{n-2} + \ldots + c_1)a = 1$$

so that $(1/c_0)(a^{n-1} + c_{n-1}a^{n-2} + \ldots + c_1)$ is an inverse for a, and thus is a unit.

==> If $c_0 = 0$ we get

$$(a^{n-1} + c_{n-1}a^{n-2} + \ldots + c_1)a = 0$$

and, since $p(x)$ is minimal for a, $a^{n-1} + \ldots + c_1 \neq 0$ and a is a divisor of zero and thus not a unit. This proves the lemma.

Proof of Proposition 10 (Chapter I)

Clearly $U(A) \subset A \cap U(B)$. Suppose $a \in A \cap U(B)$ and we will show $a \in U(A)$. Since a is a unit in B there exists $b \in B$ such that

$$ab = 1 = ba$$

It remains to show $b \in A$ (so that $a \in U(A)$).

Let $$p(x) = x^n + c_{n-1}x^{n-1} + \ldots + c_1 x + c_0$$

be the minimal polynomial for a.

Since a is a unit, $c_0 \neq 0$.

From $$p(a) = (a^{n-1} + \ldots + c_1)a - c_0$$ we get

$$(a^{n-1} + \ldots + c_1)ab = -c_0 b \qquad \text{and } ab = 1$$

so $$b = -(1/c_0)(a^{n-1} + \ldots + c_1)$$

so that b is a polynomial in a and hence $b \in A$. QED.

References

1. J. Frank Adams, Lectures on Lie Groups, Benjamin, New York 1969.

2. C. Chevalley, Theory of Lie Groups I, Princeton Univ. Press, 1946.

3. J. Helgason, Differential Geometry, Lie Groups, and Symmetric Spaces, Academic Press, 1978.

4. G. Hochschild, The Structure of Lie Groups, Holden-Day, 1965.

5. Roger Howe, Very Basic Lie Theory, American Mathematical Monthly, vol. 90 (1903), pp. 600-623.

6. S. Lang, Introduction to Differentiable Manifolds, Interscience, 1962.

7. J. P. Serre, Lie Algebras and Lie Groups, Benjamin, 1965.

8. H. Weyl, The Classical Groups, Princeton Univ. Press, 1946.

Index

Abelian group 4

Algebra 15

Algebra of matrices 15

Atlas 163

Basis for open sets 82

Binary operation 1

Bounded set 81

Cartesian product of sets 1

Cartesian product of groups 92

Category 173

Center of a group 20

Centers of Sp(1) and SO(3) 64

Center of Sp(n) 100

Centers of U(n) and SU(n) 101

Center of Spin(n) 141

Centralizer (of a set in a group) 20

Chart 163

Clifford algebra 134

Closed manifold 87

Closed set 79

Commutator 69

Commutator subgroup 69

Compact set 81

Complex numbers 9

Conjugation in $\mathbb{R}$, $\mathbb{C}$, $\mathbb{H}$ 23

Conjugate of a matrix 24

Conjugacy of maximal tori 126

Conjugate of a reflection 120

Conjugates of max T cover U(n), SU(n) 110

Conjugates of max T cover SO(n) 114

Conjugates of max T cover Sp(n) 118

Connected set 79

Continuity of a function 76

Cosets of a subgroup 67

Countability 83

Countable basis for open sets 85

Cross section 148

Curve in a vector space 35

Curve in a matrix group 36

Dense subset 124

Diffeomorphism 163

Differentiable curve 164

Differentiable manifold 164

Differentiable structure 164

Differential of a smooth homomorphism 42, 171

Dimension of a matrix group 37

Dimension of GL(n,ℝ) and GL(n,ℂ) 38, 39

Dimensions of some matrix groups 41

Direct sum of algebras 135

Discrete subgroup 94

Divisors of zero 8, 135

Eigenvector, eigenvalue 107

Eigen space 108

Exponential of a matrix 45

Field 7

Functor 174

Fundamental group of a matrix group 131

General linear groups 16

Generator (of a monogenic group) 125

Group 2

Group extensions 148

Groups of rank 1,2,3 128

Homeomorphism (of spaces) 86

Homomorphism (of groups) 4

Homomorphism of Sp(1) onto SO(3) 61

Homomorphism of Pin(k) onto O(k) 138

Indempotent matrix 33

Identity component of a group 132

Injective homomorphism 5

Inner product 23

Isomorphism (linear) 14

Isomorphism of groups 6

Isomorphism of Sp(1) and SU(2) 30

Isomorphism of Sp(2) and Spin(5) 143

Isomorphism of SU(4) and Spin(6) 143

Jacobi identity 57

Kernel of a homomorphism 19

Lattice subgroup of R^n 104

Left translation 60. 71

Length of a vector 24

Lie algebra 57, 171

Lie group 172

Lie algebras of Sp(1) and SO(3) 58

Linear map 12

Logarithm of a matrix (near I) 49

Loop group $\Omega(G)$ 131

Manifold 87

Maximal torus 95

Maximal torus in SO(n) 97

Maximal tori in U(n) and SU(n) 97, 98

Maximal torus in Sp(n) 99

Metric 74

Monogenic group 124

Nilpotent matrix 34

Normal subgroup 20

Normalizer (of a set in a group) 20

Normalizers (of max. tori) in Sp(1) and SO(3) 147

One-parameter subgroup 51

Open ball 75

Open set 78

Orthogonal groups 27

Path 80

Path in a group 130

Pin(k) 137

Primitive root of unity 22

Projection 33

Quaternions $\mathbb{H}$ 11

Quaternions have square roots 116

Quotient group 68

Rank of a matrix group 127

Reflections 31, 119

Reflections generate $\mathcal{O}(n)$ 121

Rotation group 27

Schwarz inequality 75

Simple group 129

Simply-connected group 131

Skew-Hermetian matrix 40

Skew-Symmetric matrix 39

Skew-symplectic matrix 40

Smooth homomorphism 41

Split group extension 148

Spin(k) 140

Special orthogonal group 29

Special unitary group 29

Stable subspace for a linear map 107

Subgroup 16

Subspace topology 82

Surjective homomorphism 5

Symmetric group 4

Symmetric linear map 112

Sumplectic group 27

Tangent space 37

Tangent vector 35, 165

Table of dimensions, centers, maximal tori 103

Torus 93

Trace of a matrix 54

Transpose of a matrix 24

Triangle inequality 74

Unipotent matrix 34

Uniqueness of one-parameter subgroups 53

Unit (in an algebra) 15

Unitary group 27

Universal covering group 131

Vector field 168

Weyl group 149

Supplementary Index (for Chapter 13)

Adjoint representation 190, 192

Coroot 188

Divisible Group 182

Reflection 187

Regular element 185

Root kernel 189

Roots 194

Singular element 185